Patrick Moore's Practical

For other titles published in this series, go to
http://www.springer.com/series/3192

Astronomical
Spectroscopy for
Amateurs

Ken M. Harrison

Ken M. Harrison
Cobham, UK

Additional material to this book can be downloaded from http://extras.springer.com

ISSN 1431-9756
ISBN 978-1-4419-7238-5   e-ISBN 978-1-4419-7239-2
DOI 10.1007/978-1-4419-7239-2
Springer New York Dordrecht Heidelberg London

Springer is part of Springer Science+Business Media (www.springer.com)

Preface

There have been three significant milestones in the history of observational astronomy: the invention of the telescope, photography, and the spectroscope. The development of the spectroscope has contributed more to the science than any other telescope accessory. It has been said that 85% of all astronomical discoveries have been made with the spectroscope.

Probably due to the perception that lots of mathematics and calculations are involved, plus the fact that it doesn't have the "Ohh" or "Ahh" impact of some of the spectacular astronomical images now being regularly distributed on the forums and websites, spectroscopy is an area that has been long overlooked and neglected by the amateur. By using amateur telescopes, mountings, and CCD cameras currently available, this book will show how, with the addition of a simple spectroscope we can observe and record spectra that reveal the temperature, composition, and age of stars, the nature of the glowing gases in nebulae, and even the existence of other exoplanets circling around distant stars.

The basic challenge facing the novice is where to start. What equipment will I need? Where can I find a spectroscope? How do I process the CCD image? How do I analyze my first spectrum? These questions and more are addressed in this book. Up to date information on equipment, spectroscopes, and methods available to the amateur, and more importantly "How to. . .".are all included in this book.

There are three basic sections in this book:

1. Introduction to Spectroscopy. This part provides a brief overview of the history of spectroscopy, the theory behind the spectrum lines, and types of spectroscopes.
2. Obtaining and Analyzing Spectra. Tells how to set up and use your spectroscope; describes different commercially available spectroscopes, cameras, and CCD's; explains how to analyze your spectra; and presents some interesting amateur projects.

3. Spectroscope Design and Construction. Here you will find basic spectroscope design ideas and how to construct your own spectroscope.

Each section is independent of the other, so if you want to jump straight into taking your first spectrum, go to the second section and get started!

"Spectroscope" is the generic term for visual spectroscopes, spectrographs (imaging), and spectrometers (linear CCD measuring devices). (Telescopes are not given different names when used with eyepieces, cameras, or filters, so why should spectroscopes?!)

Units of measure are always an issue of debate. Both the SI unit nanometer (nm) and Angstrom units (Å) are widely used in spectroscopy as a measure of wavelength, as are measurements and sizes in millimeters rather than inches.

As you gain practice and experience you may want to increase the resolution of your spectroscope, contribute to the ever-growing list of amateur and pro-am projects, or even construct your own spectroscope. The various sections in this book will guide you through the issues and hopefully answer your questions on all the different aspects of spectroscopy. It's a new and challenging field for amateurs, and with even the most basic of equipment it can be interesting, thought provoking, and most of all fun!

Cobham, UK   Ken M. Harrison
February 2010

Contents

Contents

About the Author

A keen amateur astronomer, Ken Harrison was born in Scotland where he trained as a mechanical engineer. He has been designing and building telescopes since the early 1960's and has built a series of spectroscopes for use on medium sized amateur telescopes. He was Section Director of the Astronomical Society of Victoria, Australia, Astrophotographic Section for 10 years and past President of the Society. Ken's university thesis (and his first publication) was *Design and Construction of the Isaac Newton 98-inch Telescope* (Strathclyde University, 1970). Since then he has published articles on optical design including "Blink Comparison" (BAA Journal Vol 87, p 94) and "Method of Radially Supporting Large Mirrors" (Vol 87, p 154). He has made contributions to the Astronomical Society of Victoria Newsletter and was for 3 years the Editor of the "N'Daba" newsletter of the Natal Centre, Astronomical Society of Southern Africa.

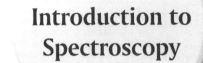

PART ONE

Introduction to Spectroscopy

CHAPTER ONE

Early Experiments in Spectroscopy

To the ancient Greeks and other philosophers around the world, light was all-pervasive and a medium connecting visible objects. Aristotle's view of light was something all bodies could have, similar to the element of fire. Early religions associated light with the Sun (as well as heat and life-giving energy) and accepted that it was a gift from the gods. Robert Grosseteste in the early thirteenth century declared light to be the "prima material," the original substance from which the universe was made, "for every natural body has in itself a celestial luminous nature and luminous fire."

The rainbow should have given philosophers cause to ponder more about the nature of light. Instead they rationalized it to be a sign from the gods. For Christians, it was God's covenant that never again would there be a flood like that experienced by Noah (see Fig. 1.1).

Al Farisi and Theodoric of Freiberg around A. D. 1300 showed how the geometry of the raindrop could produce a rainbow, but it wasn't until much later, in 1670, that Isaac Newton (1642–1727) applied his scientific reasoning to the analysis of the Sun's spectrum. He allowed a small beam of sunlight to go through his prism and produce a spectrum. With this he showed that white light is made up of many colors. The original red, yellow, green, blue, and violet were augmented by orange and indigo (to make up a "perfect seven"). ROY G BIV has become our rainbow ever since.

With his experiments in "refraction" Newton concluded that lenses would never give a color-free image (chromatic aberration) and went on to develop his reflecting telescopes as an alternative (see Fig. 1.2).

After Newton there was a hiatus in the scientific use of prisms, and it wasn't until 1750 that a Scot, Thomas Melvill (1726–1753) noted that by adding "burning spirit" to his candle, his prism showed a band of yellow light. At about the same time, John

K.M. Harrison, *Astronomical Spectroscopy for Amateurs*, Patrick Moore's Practical
Astronomy Series, DOI 10.1007/978-1-4419-7239-2_1,
© Springer Science+Business Media, LLC 2011

Figure 1.1. *A Lesson In The Rainbow,* late fourteenth-century manuscript of "De proprietatibus rerum." (On the Property of Things) by Bartholomaeus Anglicus.

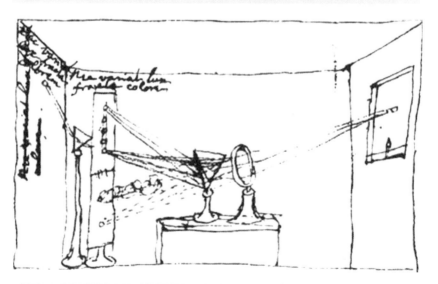

Figure 1.2. Newton's prism experiments. (Courtesy New College Library.)

Dollond (1706–1761), an English optician, appreciated that not all glass dispersed light by the same amount; he found (in 1758) that by combining a clear crown glass and a leaded flint glass that the chromatic aberrations could be significantly reduced. Here we have the first evidence of appreciation that the spectrum produced by different types of glass was different enough to be useful.

There is very little recorded evidence of other spectral experiments in the 1700's until William Herschel (1738–1822) was finally persuaded by his colleague William Watson (1715–1787) to experiment with a prism to view the spectra of the brighter stars. In 1797 he recorded the distribution and intensity of the various colors in the spectra, but did not pursue the interest.

William Hyde Wollaston (1766–1828) improved his view of the solar spectrum by adding an "elongated crevice, 1/20″ wide" between the prism and the Sun. This showed him seven distinct dark lines in the spectrum (1802); these he associated with the "natural boundaries" between the colors. Such was the state of knowledge in the early 1800's.

Fraunhofer Lines

Label	Wavelength (nm)	Source
A-band	759.4–762.1	Atmospheric O_2
B-band	686.7–688.4	Atmospheric O_2
C	656.3	$H\alpha$
a-band	627.6–628.7	
D1	589.6	Na I
D2	589.0	Na I
E	526.9	Fe I
b1	518.4	Mg I
b2	517.3	Mg I
c	495.8	Fe I
F	486.1	$H\beta$
d	466.8	
e	438.3	Fe I
f	434.0	$H\gamma$
G	430.8	Ca/Fe (430.77/430.79)
g	422.7	Ca I
h	410.2	$H\delta$
H	396.8	Ca II
K	393.3	Ca II

Josef von Fraunhofer (1787–1826) took up the challenge to measure the dispersion of the glass he was producing. At this time there was very little scientific analysis carried out in the manufacture of glass; making glass was an art, not a science. There was a degree of trial and error every time!

By making small prisms from each batch of glass and measuring and recording the dispersion, Fraunhofer could quickly ensure that the process was giving the outcome he needed. This allowed him to manufacture the best objectives of the time, and his large telescopes were the envy of all astronomers. He designed and built the 9.5″ Great Dorpat Refractor in 1824, mounting it on his new "German Equatorial mount."

To improve his measuring device, which he called a "spectroscope," he added a slit in front of the collimating lens; this gave a cleaner image of the spectrum. He also noticed that sunlight, when dispersed by his prisms, always seemed to have dark lines in the same positions, and eventually, in 1815, he mapped some 324 lines; the more prominent ones are still called "Fraunhofer lines". See Fig. 1.3. Although he didn't understand how these lines were produced, he made good use of them as "standard wavelengths" for his optical testing.

Figure 1.3. Fraunhofer's spectrum. (Photo courtesy of NASA.)

Fraunhofer also placed a prism in front of his telescope objective (in 1823) and observed the stars and planets. He noted that similar lines appeared in the stellar spectra. This very first objective prism is currently on display in the Deutsches Museum in Munich. See Fig. 1.4.

To further improve his spectral measurements he developed the diffraction grating. Early examples were made by winding fine wire around a frame. With these

Figure 1.4. Fraunhofer's early objective prism. (From Dr. L. Ambronn, "Handbuch der Astronomischen Instrumentenkunde- II Band",1899.)

gratings, which gave the spectrum a more uniform scale, it was easier to accurately measure the position of the various dark lines.

It can therefore be said that Fraunhofer was certainly the father of spectroscopy. He invented and developed the spectroscope to the stage that it could be used as a scientific tool to view and measure spectra.

Giovanni Amici (1786–1863), director of the Florence Observatory in Italy, employed a combination of crown and flint prisms in his dispersion prism arrangement that effectively gave a straight through, undeviated spectrum. This "Amici prism" design was widely used in subsequent Direct Vision star spectroscopes.

It was in 1859 that Gustav Kirchhoff (1824–1887) and Robert Bunsen (1811–1899) finally solved the problem of the dark (and bright) lines in the spectrum. Their experiments showed conclusively that various elements gave rise to their own unique bands (always in the same position) in the spectrum. With this connection now established many chemists took up the challenge and established a database of element lines.

Kirchhoff, in furthering his investigations, developed his three laws:

(1) An incandescent solid or a gas under high pressure will produce a continuum spectrum.
(2) An incandescent gas under low pressure will produce an emission-line spectrum.
(3) A continuous spectrum viewed through a low-density gas at low temperature will produce an absorption-line spectrum.

The era of scientific research had begun. No more rainbows!

For Further Reading

Abbott, D. (Ed.), *The Biographical Dictionary of Scientists-Astronomers.* Blond Educational (1984).

Web Pages

http://home.vicnet.net.au/~colmusic/clario2.htm

CHAPTER TWO

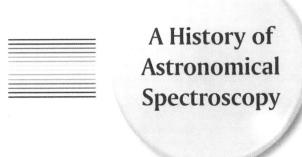

A History of Astronomical Spectroscopy

The publication of Kirchhoff's and Bunsen's work brought the awareness of the spectroscope, and what it could reveal, to a wider audience, including astronomers. The obvious question was, how could this new instrument be used to analyze the light from the Sun and stars?

Auguste Comte (1798–1857), a French philosopher stated this in 1835: "We may in time ascertain the mean temperature of the heavenly bodies: but I regard this order of facts as for ever excluded from our recognition. We can never learn their internal constitution, nor, in regard to some of them, how heat is absorbed by their atmosphere." He was about to be proved wrong!

One of the first astronomers to apply the spectroscope to his telescope was William Huggins (1824–1910), an English amateur. To quote from his later book:

> I soon became a little dissatisfied with the routine character of ordinary astronomical work, and in a vague way sought about in my mind for the possibility of research upon the heavens in a new direction or by new methods. It was just at this time ... that the news reached me of Kirchhoff's great discovery of the true nature and the chemical constitution of the sun from his interpretation of the Fraunhofer lines.

> This news was to me like the coming upon a spring of water in a dry and thirsty land. Here at last presented itself the very order of work for which in an indefinite way I was looking – namely, to extend his novel methods of research upon the sun to the other heavenly bodies. A feeling as of inspiration seized me: I felt as if I had it now in my power to lift a veil which had never before been lifted; as if a key had been put into my hands which would unlock a door which had been regarded as for ever closed to man – the veil and the door behind which lay the unknown mystery of the true nature of the heavenly bodies.

K.M. Harrison, *Astronomical Spectroscopy for Amateurs*, Patrick Moore's Practical
Astronomy Series, DOI 10.1007/978-1-4419-7239-2_2,
© Springer Science+Business Media, LLC 2011

For the next 40 years he and his wife Margaret dedicated their time and resources to observing the sky with the spectroscope.

Huggins designed and built all his prism spectroscopes and pioneered new techniques such as providing a reference spectrum from an electric spark and a reflection slit to improve guiding the spectroscope on a star. See Fig. 2.1.

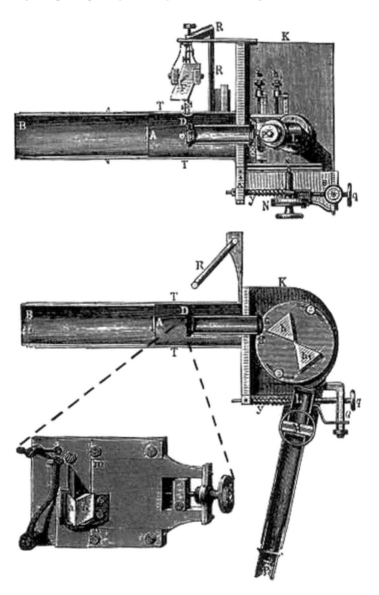

Figure 2.1. Huggins spectroscope. (WIKI.)

From his observatory at Tulsa Hill just outside London, Huggins was the first to observe emission lines in the spectra of nebulae; he also correctly applied Doppler's principle to his spectra to determine the radial velocity of a star (Sirius). In 1869, Huggins developed a technique to allow the observation of solar prominences without the need for a solar eclipse. He later also correctly identified the ultraviolet lines of hydrogen on photographic plates.

Another active observer at the time was Father Angelo Secchi (1818–1878) of the Vatican Observatory. Secchi observed the spectra from over 4,000 stars and developed a stellar classification system that was used for almost 50 years.

In 1863 he announced his Class I (strong hydrogen lines) and Class II (weaker hydrogen lines with numerous metallic lines) stars; by 1866 he had added Class III (bands stronger towards the blue, plus metallic lines), and in 1868 Class IV (deep red stars with bands opposite to Class III). He later added Class V (emission spectra). See Fig. 2.2.

The Mertz 12° objective prism 162 mm diameter (made in 1872) used by Secchi for his later research, was displayed at the 2009 ASTRUM exhibition in Rome.

Anders Jonas Angstrom (1814–1874), using an early grating spectroscope, mapped the solar spectrum with greater accuracy than had been done previously. In 1868 he published an atlas of over 1,000 lines, their positions recorded in units of 10^{-10} m. This is now known as Angstrom Units (Å) and is still widely used.

Henry Draper (1837–1882) succeeded in obtaining a photograph of the spectrum of Vega, which clearly showed the hydrogen absorption lines (1872). The advent of the dry photographic plate in the mid-1880s allowed early investigators to carry out the long exposures required to record spectra (Harvard Observatory obtained the first spectrum of a meteor in 1897). The use of these dry plates and later photographic film heralded the beginning of the transfer of spectroscopy from amateurs to professional astronomers.

With monies from the Henry Draper Memorial Fund, Edward Pickering (1846–1919) and his team at Harvard Observatory followed the work of Secchi in recording and cataloging stellar spectra. Using an objective prism mounted in front of the telescope objective he was able to quickly amass large amounts of low resolution spectra (He used objective prisms with angles from 5 to 7° mounted on telescopes up to 13″ aperture to obtain the spectra). The subsequent "Henry Draper Catalogue" of stellar spectra was based on separate classes and sub classes; W O B A F G K M. Updated and enhanced versions of this catalog are still used today.

The work of Henry Rowland (1848–1901) in perfecting his grating ruling engine in 1882 allowed the production of large diffraction gratings that gradually took over from prisms in professional spectroscopes.

By the turn of the century the era of the amateur scientist was drawing to a close; larger and larger spectroscopes and telescopes were producing scientific results that would determine the direction of astrophysics for the next 100 years.

The interested amateur could acquire spectroscopes made by John Browning (1835–1925). These were small direct-vision Amici prism instruments (D-V) for stellar observing, and they established the trend for the next 40 years. Being a dedicated visual instrument, the results were limited to viewing spectra of the Sun, brighter stars, and nebulae.

VISUAL SPECTRA TYPE V COMPARED WITH TYPE IV (VOGEL).
1. V Can. Ven. ; 2. XVIIIb 3m, S. 21' 16' ; 3. VIb 51m, S. 23' 42' ; 4, 5, 6. Wolf-Rayet, Nos. 1, 2, 3. Cygnus.

Figure 2.2. Visual spectra of Secchi Type IV and V by Vogel. (From Preface to Webb's Celestial Objects for Common Telescopes, 1917.)

Commercial D-V instruments continued to be produced by Adam Hilger, Ltd., among others and examples by GOTO (Japan) and LaFeyette (USA) were widely used by amateur astronomers in the 1950's and 1960's (see Fig. 2.3).

Figure 2.3. Various D-V spectroscopes. From the top: John Browning. (circa 1880), Adam Hilger (circa 1920), LaFeyette (1960), GOTO (1970), Meiji-Labax (1980), and Surplus Shed (2004).

By the 1970's, transmission gratings became more readily available to amateurs, and these were used to construct spectroscopes capable of much more serious work than the early D-V instruments. This trend has continued, and nowadays instruments are being constructed with reflection gratings that can give spectral resolutions capable of measuring Doppler shifts and spectroscopic binary stars.

Currently (2010) there are at least four manufacturers supplying spectroscopes for the amateur (See later for details).

Further Reading

Clerke, A. M. *A Popular History of Astronomy*. Adam & Charles Black (1887).

Huggins, W. *The Scientific Papers of Sir William Huggins*. Wesley & Sons (1909).

Jackson, M. W. *Spectrum of Belief*. Joseph Von Fraunhofer and the Craft of Precision Optics. MIT Press (2000).

Maunder, E. W. *Sir William Huggins and Spectroscopic Astronomy*. TC&EC Jack (1913).

Young, C. A. *The Sun*. Kegan Paul Trench & Co (1882).

Web Pages

http://messier.obspm.fr/xtra/Bios/huggins.html

http://www.fraunhofer.de/en/about-fraunhofer/joseph-von-fraunhofer/

http://gallica.bnf.fr/ark:/12148/bpt6k30204.image.f623

CHAPTER THREE

Theory of Spectra

Kirchhoff's Laws

The application of Kirchhoff's laws to astronomical spectra allowed Huggins and others to understand the nature of stellar atmospheres. Here are the laws:

1. An incandescent solid or a gas under high pressure will produce a continuum spectrum.
2. An incandescent gas under low pressure will produce an emission-line spectrum.
3. A continuous spectrum viewed through a low-density gas at low temperature will produce an absorption-line spectrum.

Stars generally produce a continuum spectrum due to their gaseous nature, and the surrounding stellar atmosphere shows the various elements as absorption lines. This then allows the astronomers not only to determine the elements in the star but their concentrations.

Emission spectra are seen when a gaseous nebula is excited by nearby stars and provide a unique fingerprint of the gases, i.e., hydrogen, oxygen, etc. Some stars, notably Be stars, also show emission features due to a local transfer of matter to a surrounding darker dust/gas cloud.

K.M. Harrison, *Astronomical Spectroscopy for Amateurs*, Patrick Moore's Practical
Astronomy Series, DOI 10.1007/978-1-4419-7239-2_3,
© Springer Science+Business Media, LLC 2011

Black Body Radiation

Kirchhoff also proposed (1860) his black body theory of thermal emission to demonstrate the thermal properties of a body in equilibrium. The distribution of light emitted by a body will follow a regular shape dependent on its temperature. This is the black body curve.

When a body is heated, the peak wavelength of the light emitted varies, peaking initially in the infrared and moving towards the UV. A dull red is normally visible when material (e.g., steel) is heated to around 800°; as the temperature increases, the color becomes orange and eventually white (white hot). Wein and Planck later developed the mathematical model for this transition, which can be simply stated as:

$$\lambda_{max} = 2,897,820/T$$

where T is the temperature in degrees Kelvin, and λ_{max} the peak emission wavelength in nm. See Fig. 3.1.

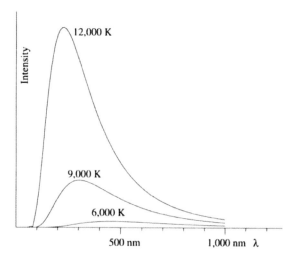

Figure 3.1. Planck curves. (from H. Karttunen et al. *Fundamental Astronomy,* 4th Ed. Springer, 2003.)

The shape of the continuum spectrum reflects the surface temperature of the star. It is assumed that stars, etc., act like black bodies The Planck curve shows that the position of the peak energy moves towards the blue/UV region with increasing temperature. The Sun, a 6,000° star, peaks in the green, around 500 nm (see Fig. 3.2), whereas a hotter star, like Vega, peaks in the blue around 400 nm. The shape of the spectral continuum can therefore assist in determining the surface temperature of

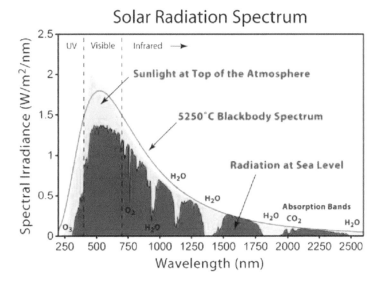

Figure 3.2. Solar radiation and atmospheric absorption. (WIKI.)

the stars. See Fig. 3.2. This, together with information of element absorptions (See Stellar Spectra, below) can allow the accurate classification of a star to be determined.

Quantum Theory

In amateur astronomical spectroscopy we are constantly faced with the challenge of analyzing the spectra we obtain. Ideally we want to determine which absorption or emission feature belongs to which element. The plate scale of the available reference spectra can vary and make identification difficult. It helps to understand how the various lines are formed, and this can sometimes give a clue as to which lines may or may not be visible.

As per Kirchhoff's laws, the emission spectra of the various stellar elements merge together to give a continuous bright spectrum and only the elements in the surrounding stellar atmosphere give us the absorption and emission features.

Each element will absorb/emit in a unique series of wavelengths dependent on temperature. These lines can also be modified by pressure and electromagnetic fields. See Fig. 3.3.

Quantum mechanics determine the various energy levels available for atomic electrons, and depending on movement of these electrons, the energy difference gives rise to a number of photons, which are either absorbed or emitted.

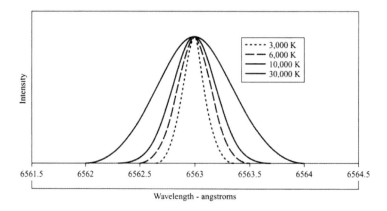

Figure 3.3. Broadening of the Hα line due to temperature. (K. Robinson.)

Energy and wavelength are related by Planck's constant:

$$\Delta E = h\nu$$

Where ΔE is the change in energy, ν, the frequency of the radiation (photon) emitted/absorbed, and h, Planck's constant $= 6.6256 \times 10^{-34}$ Js.

When energy is added to the atom – through heat, collision, electromagnetic fields, etc. – the electrons move to stable orbits further from the nucleus. In absorbing this energy, the photon equivalent is seen as dark absorption lines in the spectrum. Similarly, when photons are emitted during energy loss, emission lines are seen.

The various energy levels of the electron movements can be shown schematically in Grotrian diagrams. See Fig. 3.4.

As hydrogen only has one electron, the available energy levels can easily be mapped, and the Balmer series (see later) represents these transitions in the spectrum.

Forbidden Lines

When an element absorbs or emits significant energies, the electrons can be moved through many of the available orbits or even displaced from the atom completely. The atom is then said to be ionized. When more than one electron is removed, we can get double or even triple ionization. Some of these transitions have a very low probability of occurrence. As they are seldom seen in the laboratory this gave rise to the terminology "forbidden lines."

The symbols used in spectroscopy for ionized elements varies from those used by chemists. Chemists would use a multiple sign convention (Ca^+, O^{++} etc), whereas

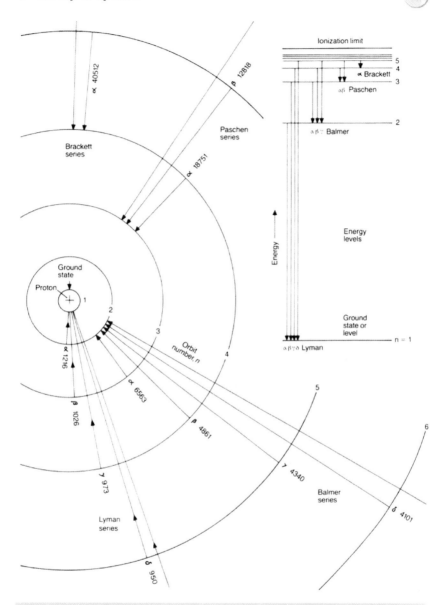

Figure 3.4. Balmer series. (From J.B. Kaler's *Stars and their Spectra,* Cambridge, 1989.)

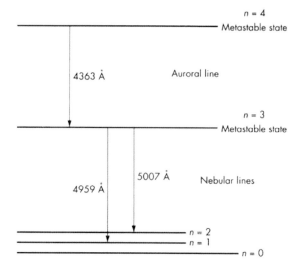

Figure 3.5. OIII ionization diagram. (K. Robinson.)

in spectroscopy we use square brackets and the uppercase Roman numeral I, where the base state is I and subsequent ionization levels are [CaII] [OIII], etc. See Fig. 3.5.

As the ionization energy levels required were not commonly available in the laboratories of the 1800's the early spectroscopists could not replicate the necessary conditions to see sufficient ionized material and therefore did not recognize the emission/absorption "signature" of the ionized element. This gave rise to "new" elements being named, i.e., "Nebulium" for [OIII] and "Coronium" for highly ionized metals.

For further reading and more detailed explanation of the various atomic structures, additional line series (Lyman, Paschen, etc.) see the Further Reading list at the end of this chapter.

Doppler and Red Shifts

Christian Doppler (1803–1853) investigated the change in sound frequency caused by velocity changes between a source and an observer. This theory was then extended to visible light. Doppler shift is visible in the spectrum when there is relative movement between the observer and the object. As the object recedes the visible wavelengths are moved towards the red. The effect can be seen in the rotation of the Sun and planets as well as binary stars. It allows the measurement of radial velocities only, i.e., towards or away from the observer.

In the case of distant galaxies, quasars, etc., the speed of recession (Redshift = z) can be significant, and velocities near the speed of light can give rise to significant movements in the spectral lines.

$$\frac{\Delta\lambda}{\lambda} = \frac{(1 + (v\cos\theta/c))}{\sqrt{(1 - (v^2/c^2))}} = z$$

where $\Delta\lambda$ is the shift in the spectral line, λ its stationary wavelength, v the velocity of recession (or approach), and c the velocity of light (2.998×10^8 ms^{-1}). The line of sight angle, θ, for pure radial velocities =0.

It is possible with amateur-sized spectroscopes to measure this shift. The shift in spectral lines due to Doppler/redshift is:

$$\Delta\lambda/\lambda = v/c$$

where $\Delta\lambda$ is the shift in the spectral line, λ its stationary wavelength, v the velocity of recession (or approach), and c the velocity of light.

To measure radial velocities of 10 km/s would require a spectral resolution of approx 1 nm/mm (10 Å/mm) and a suitable range of reference lines for comparison. Objective prisms are not normally effective for this type of work due to the low resolution and lack of comparative spectral lines. Transmission gratings have been successfully used to record the redshift of quasars by using the zero image and other "stationary" star spectra as reference.

Solar Spectrum

The solar spectrum can be observed with most spectroscopes. In the absence of a slit, a polished reflector can be used, i.e., a needle.

[CAUTION: Never look directly at the Sun with – or – without optical aid.]

There are many good references that provide detailed, annotated solar spectra. These generally cover the visible wavelengths from 380 nm through to the far red at 750 nm, with resolutions down to 0.01 nm (0.1 Å).

The sodium doublet at 589.0/589.6 nm makes a good test for low resolution spectroscopes, and picking out the "magnificent seven" Fe lines between them demonstrates the performance of R = 5000 instruments.

Another easy test is the magnesium triplet in the green, 430.0/431.0/431.2 nm.

Telluric lines of atmospheric oxygen and H_2O bands are evident at the extreme red end of the visible the spectrum, just beyond the wide and obvious Hα line at 656.3 nm. At sunset and sunrise these bands become more obvious due to the longer atmospheric path (Noting their positions will help when analyzing stellar spectra, as they obviously occur in them as well).

Huggins used his slit spectroscope in 1869 to record the first observation of Hα prominences without the benefit of a solar eclipse. This can be easy replicated today, and it is possible to image both prominences and surface detail in any wavelength (i.e., CaK, Hα etc) with a suitable instrument.

Stellar Spectra

Secchi's five classifications of stellar spectra didn't survive into the twentieth century. Following Henry Draper's premature death in 1882 his widow, Anna Mary, established the Henry Draper Memorial Fund. This fund was subsequently used by Harvard College Observatory under E. C Pickering to produce the Henry Draper Catalogue (HD) based on the spectra of over 230,000 stars. The catalog quickly became the astronomical standard. Pickering used various telescopes and objective gratings to collect the stellar images that were analyzed by Mrs. Williamina Paton Fleming (1857–1911) (who also discovered the Horsehead Nebula in 1888) and later Antonia Maury (1866–1952), and finally Annie Jump Canon (1863–1941).

The original HD stellar classifications were based on the observed strength of the hydrogen lines, and an alphabetical sequence was used, i.e., A, B, C, etc. This sequence generally followed the temperature of the stars. As the instruments and spectroscopes improved, it became obvious that some of the classifications were either incorrect or out of sequence. Classes C, E, and H were dropped. Maury developed a further sequence identifying the stars in 22 groups using Roman numerals. This was not readily accepted, and it was left to Cannon to come up with the sequence we are familiar with today:

$$\begin{array}{cc} \textbf{W} & \textbf{C} \\ | & | \\ \textbf{O-B-A-F-G-K-M} \\ & | \\ & \textbf{S} \end{array}$$

The R and N stars (carbon-rich stars) have been combined into the C class sitting of the G branch and S stars (Zirconium) sitting of the M branch. The final classifications of P (planetary nebulae) and Q (nova) are seldom used today.

Canon herself proposed the mnemonic "Oh be a fine girl kiss me" to remember the classification sequence. Nowadays a mnemonic-like: "Only Bored Astronomers Find Gratification Knowing Mnemonics" may be more acceptable!

The H-R Diagram

Ejnar Hertzsprung (1873–1967) and Henry Norris Russell (1877–1957) noted that when the spectral classifications were plotted against the absolute visual magnitudes that the graph obtained showed a grouping of stars in various regions. This "H-R Diagram" clearly showed the differences between giant stars, a group sitting above a larger continuous band of stars (main sequence), and the dwarf stars, sitting below. See Fig. 3.6.

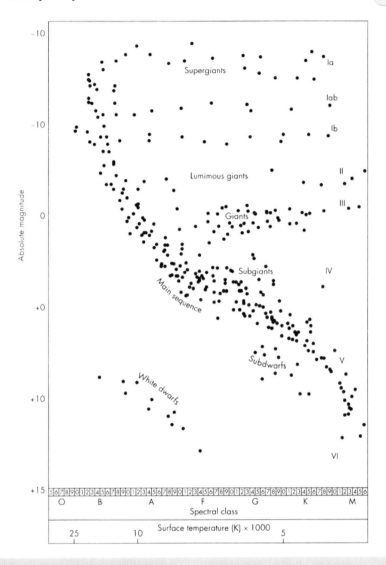

Figure 3.6. HR diagram. (S. Tonkins.)

HD Classifications

In the early 1940s the classification sequence was further refined by William Morgan (1906–1994), Philip Keenan (1908–2000), and Edith Kellman (1911–2007) of Yerkes Observatory into the MKK system. This resulted in the rationalizing of the subseries (generally 0–9) and the addition of luminosity classes to differentiate between giant, main sequence and dwarf stars and a series of prefixes and suffixes to give further information.

Based on objective prism spectra the following criteria were used for classification:

Class O   Ionized helium, silicon, nitrogen, etc.
Class B   Neutral helium, stronger Balmer lines of hydrogen.
Class A   Strong Balmer lines, appearance of ionized calcium, and neutral metals.
Class F   Weaker Balmer lines and ionized metals; strong neutral metal lines.
Class G   Balmer lines very weak, strong ionized calcium lines, and stronger neutral metal lines.
Class K   Molecular bands (TiO) appear; neutral metal lines
Class M   Dominated by molecular bands (TiO); strong neutral calcium lines.

The Spectral Sequence

O4–O9.5
B0–B9.5   (B0.5 added, B4, B6, and B9 omitted)
A0–A7   (omitting A1, A4, and A6)
F0–F9   (omitting F1, F4, and F6)
G0–G8   (omitting G1, G3, G4, G6, and G7)
K0–K5   (omitting K1)
M0–M9   (omitting M5 and M6)

Luminosity Classes

0 –   Extremely bright supergiants
Ia-   Bright supergiants
Ib-   Supergiants
II-   Bright giants
III-   Normal giants
IV-   Subgiants
V-   Main sequence stars
VI, sd-   Subdwarfs
VII, wd-   White dwarfs

Suffixes and Prefixes Used in Spectral Classifications
Prefix

b-   wide lines
a-   normal lines
c-   very narrow lines
d-   dwarf (main sequence, luminosity class V)
g-   giant (luminosity class III)
sd-   subdwarf (luminosity class VI)
wd-   white dwarf (luminosity class VII)

Suffix

e-	emission lines
em-	emission lines (metallic)
er-	reversed emission lines
ep-	peculiar emission lines
eq-	P cygni type emissions
f-	He and N emissions (O type stars)
n-	diffuse lines
nn-	very broad diffuse lines
s-	narrow or sharp lines
k-	interstellar lines
v-	spectrum is variable
pec-	peculiar spectrum
m-	strong metallic lines
wk-	weak lines
!-	marked characteristics

The spectrum of a star can then be classified and recorded as a combination of the above.

Our Sun is a G2V star; i.e., a yellow main sequence star in the G2 class with a surface temperature of 6,000°C. Sirius, on the other hand, is a bright white A1V main sequence star and Arcturus a normal orange giant K1 III star.

Standard Spectral Lines and Reference Spectra

To accurately calibrate the spectrum we need a series of reference lines that can be used to measure the dispersion of the spectroscope and its resolution (For spectroscopes without slits, i.e., small transmission gratings or objective prisms, calibration lamps cannot be used and other methods must be used, for example, comparison spectra, off-set from a zero image).

Once we have calibrated our spectrum it is possible to measure the positions of absorption and emission lines to determine the atoms or molecules involved.

The professionals make use of special reference lamps, such as thorium/argon, which emit their light in well-documented discrete lines. For the average amateur these are very expensive, and alternative methods need to be considered.

The easiest method of spectral calibration is to use known lines in the spectrum itself. The most obvious of these are the Balmer series of hydrogen, formulated by Johann Jakob Balmer (1825–1898) in 1885. See Fig. 3.7.

The formula for the series is:

$$1/\lambda = 109677.1\,(1/n^2 - 1/m^2)$$

where n = 2 for the Balmer series and m is an integer > n.

Balmer Series

m=3 Hα	656.3 nm
m=4 Hβ	486.1 nm
m=5 Hγ	434.0 nm
m=6 Hδ	410.2 nm
m=7 etc	397.0 nm
m=8 etc	388.9 nm
m=9 etc	385.5 nm
m=10 etc	379.7 nm

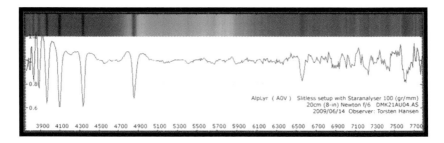

Figure 3.7. Spectrum of Vega. (T. Hansen)

Where the spectroscope has an entrance slit, a reference lamp can be used to illuminate the slit and provide a series of reference lines. Neon indicator bulbs can be used as a reference light source.

The neon gas gives a series of bright emission lines covering 585.25 nm through to 724.52 nm. See Fig. 3.8 below.

Flouro Spectrum

As an alternative to the neon, an energy-saving fluoro bulb such as the Osram Dulux Star can be used. These flouro bulbs provide an excellent series of mercury lines from 435.83 to 587.4 nm. There are also some strong emission lines at 611.3 and 631.1 nm. See Fig. 3.9.

Other Useful Spectral Reference Lines

Sodium doublet 589.0 and 589.6 nm
Nebula emission lines

NII	658.4 nm
Hα	656.3 nm

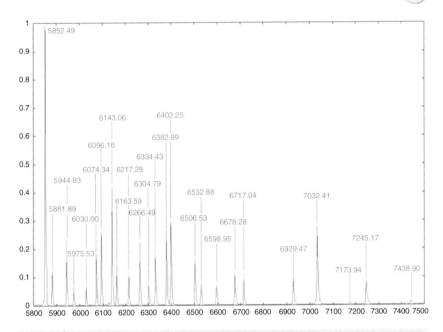

Figure 3.8. Neon lamp spectrum. (C. Buil.)

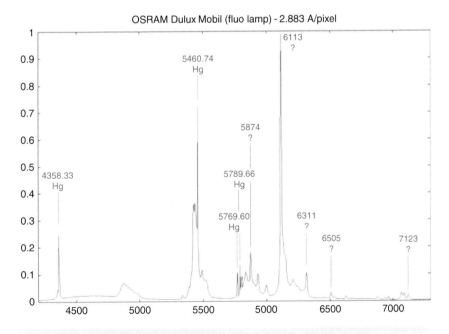

Figure 3.9. Fluoro lamp spectrum. (C. Buil.)

OIII	500.7 nm
OIII	495.9 nm
Hβ	486.1 nm
HeII	468.6 nm
Hγ	434.0 nm
NeIII	386.9 nm

O_2 and H_2O atmospheric telluric lines

H_2O	594.45 nm
H_2O	595.86 nm
H_2O	596.83 nm
O_2	627.8 nm
O_2	686.9–694.4 nm
O_2	760.6 nm

Airy Disk, Rayleigh Limits, and FWHM, PSF

Before considering the optics used in a spectroscope, we need to familiarize ourselves with the basics of diffraction and resolving power.

As we know, stars are so far away that they appear as pinpoints of light. When we observe a star through a telescope at high magnification, it's immediately obvious that the star image appears as a small disk of light surrounded by faint rings rather than a discrete point. So what turns this infinitely small pinpoint of light into a shimmering disk?

George Biddell Airy (1801–1892) in the late 1820's investigated the effects of diffraction and found that a perfect star image would form a disk (now called an Airy disk) brighter in the middle, surrounded by a series of concentric rings. Some 80% of the light is concentrated in the disk, the rest being spread out into the various rings. See Figs. 3.10 and 3.11.

The size of the disk is defined by the position of the first dark minimum between the disk and the first ring. The radius of this disk is defined as:

$$\text{Sin}\theta = 1.22\lambda/D$$

For small angles $\text{Sin}\theta \approx \theta$ (radians), where λ is the wavelength of the light and D the diameter of the objective.

We can see then that the angular size of this disk will vary with telescope aperture. The larger the telescope, the smaller the disk size. The linear size of the disk in microns (μm) using a telescope with a focal length, F, is

$$\text{Airy disk diameter} = 2.44\lambda F/D \, (\mu m)$$

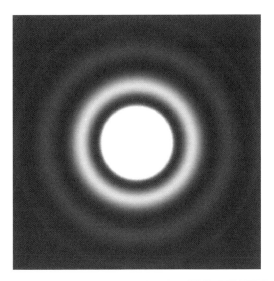

Figure 3.10. Airy disk.

A 200 mm f8 telescope would give a diffraction disk of 11 μm, whereas it would be 5.5 μm for a 400 mm f4 system. The 400 mm aperture would collect almost four times as much light and concentrate it in a disk four times as small, giving sixteen times the intensity. Optics that show a clean, undistorted Airy disk are said to be "diffraction limited."

Lord Rayleigh (1842–1919) went on to show that the theoretical resolution of a telescope could be defined by the size of the Airy disk, and stated that when the center of one disk lay in the first minimum of another they could be seen as separate stars (Figs. 3.11c and 3.12).

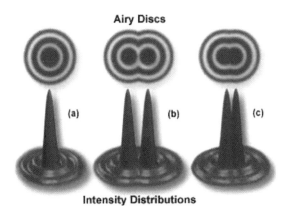

Figure 3.11. Airy disk intensity distribution. (Courtesy Olympus.)

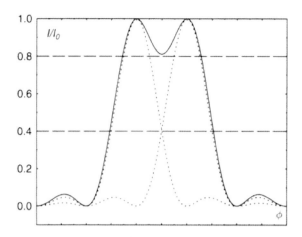

Figure 3.12. Diffraction profiles – rayleigh limit. (From *Spectrophysics,* Springer 1999.)

    The intensity distribution in the Airy disk approximates to a Gaussian curve, and 40% of the light is contained within 50% of its diameter.

    The term "Full Width Half Maximum" (FWHM) is widely used as a measure of the image size based on this Gaussian intensity distribution. See Fig. 3.13.

    Mathematically it equals:

$$2\sqrt{(2\ \ln 2)}\ \sigma \approx 2.355\sigma$$

where σ is the standard deviation of the curve.

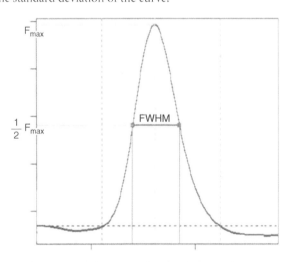

Figure 3.13. FWHM of an intensity curve.

Based on the Rayleigh criteria for resolution, the FWHM sizes of two curves when touching gives a resolved image. So, we can use the FWHM as a measure of the resolving power of an optical system.

The diffraction images produced by an optical system are defined in terms of the Point Spread Function (PSF). This basically provides a mathematical "conversion" equation between the ideal and real world outcomes.

A good example is the Airy disk itself. The "real" image is an infinitely small pinpoint of light, but it's recorded by our telescopes as a measurable disk of light.

The same thing happens in a spectroscope. An emission line, say from neon, has a minute bandwidth, maybe <0.001 nm wide – effectively a very narrow line (see Fig. 3.14a).

When recorded by a spectroscope even with perfect optics and no aberrations, it will still appear as a broad line with an intensity curve similar to the Airy disk (see Fig. 3.14b).

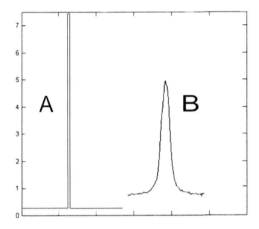

Figure 3.14. Line intensity vs. actual profile.

The instrument resolution can therefore be measured in terms of the FWHM of this profile.

Focus and Exit Pupil

Getting a spectrum and/or star image accurately in focus is important in obtaining the highest resolution with a spectroscope. Slit-less spectroscopes such as the Star Analyser and Rainbow Optics grating rely on precise focus to give the sharpest spectral image. In spectroscopes with a slit the overall efficiency is dependent on the star image being focused on the entrance slit to give a collimated beam to the grating. Likewise the imaging camera lens must allow the image of the spectrum to be brought to a tight focus.

The depth of focus, that is, the allowable focusing error, is very much dependent on the f/ratio of the objective. Slower objectives/camera lens have a much greater depth of image focus than faster lenses. See Fig. 3.15.

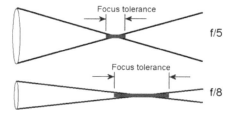

Figure 3.15. Focus tolerance vs. focal ratio.

There are various methods of calculating the maximum allowable out of focus. Sedgwick gave $\Delta f = 4(1.22\lambda F^2)$ based on his definition of the "circle of least confusion." Some astro-photographic books talk about the "critical focus zone" (CFZ) and give $\Delta f = 2.2F^2$, based on perfectly collimated images. If we work to the Raleigh limit criteria of $\frac{1}{4}$ wave accuracy we find the allowable out of focus is approx. $\Delta f = 4\lambda F^2$ where F is the focal ratio of the objective/camera lens and λ the wavelength of the light being considered. Straightaway you can see that the focus tolerance in blue light will be half that of red light (350/700 nm). Substituting some common f/ratios and working with green light (550 nm) gives the results shown in Table 3.1 below.

Table 3.1 Focal ratio vs. focus tolerance

Focal ratio (f)	Focus tolerance (mm)
10	0.220
8	0.141
5	0.055
4	0.035

If we have a Crayford-type focuser it may have a travel of about 18 mm per revolution of the focuser knob. This means that to achieve good focus with an f10 objective we need to be able to accurately rotate the knob within $(0.22/18 = 0.012)$ of a revolution, or 4.4°; for an f5 system this becomes just 1.0°!!

Motorized focusers and/or supplementary gearing or helical focusers can assist, but the challenge remains.

Exit Pupil and Eye Relief

When a direct vision spectroscope (or prism/filter grating) is used behind an eyepiece the image of the star being observed is contained in the exit pupil of the eyepiece. This should not be confused with the eye-lens diameter of the eyepiece, which can be up to 25 mm or so. The exit pupil (or Ramsden disk) of an eyepiece is the size of the parallel emerging light beam and represents the diameter of the objective divided by the magnification of the eyepiece. For example, if a 25 mm focal length eyepiece is used on a 100 mm f5 telescope, then the magnification is (100 * 5)/25 = ×20. At this magnification the exit pupil will be 100/20 = 5 mm diameter. As the light exits from the whole of the eye-lens surface, this exit pupil is produced by the various off-axis beams, where they intersect at a point behind the eyepiece; the distance from the eye-lens to this image position is called the eye relief of the eyepiece. To see the full field of view, the eye should be positioned at this point. See Fig. 3.16.

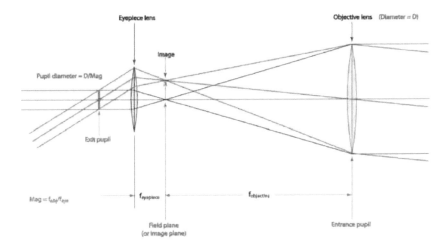

Figure 3.16. Exit pupil and eye relief. (From Al Harper, Yerkes.)

As far as the spectroscope/prism is concerned the starlight forms a collimated beam that can be dispersed and focused to form a spectrum.

Further Reading

Gray, R.O., Corbally, C. J. *Stellar Spectral Classification*. Princeton University Press (2009).
Kaler, J. B. *Stars and their Spectra*. Cambridge University press (1989).
Robinson, K. *Spectroscopy: The Key to the Stars*. Springer (2007).
Sedgwick, J. B. *Amateur Astronomer's Handbook*. Enslow Publishers (1980).
Suiter, H. R. *Star Testing Astronomical Telescopes*. Willmann Bell (2001).

Tennyson, J. *Astronomical Spectroscopy.* Imperial College Press (2005).
Wodaski, R. *The New CCD Astronomy.* New Astronomy Press (2002).

Web Pages

http://astrosurf.com/buil/us/spe2/hresol4.htm
http://spiff.rit.edu/classes/phys301/lectures/comp/comp.html

CHAPTER FOUR

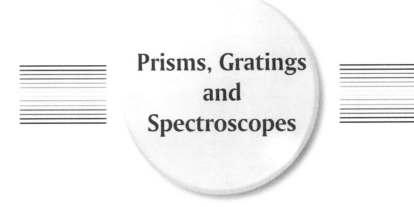

Prisms, Gratings and Spectroscopes

Spectroscopes come in all shapes and sizes, a bit like telescopes. Some are designed to be better for obtaining stellar spectra, others are made for solar and extended objects such as planets, nebulae, and deep sky objects (DSO's). Spectroscopes are based on using prisms, transmission gratings, or reflection gratings, and depending on their use, they may or may not include an entrance slit.

Dispersion, Plate Scale, and Resolution

All spectroscopes are designed to separate light into various wavelengths and spread it into a colored band called a spectrum. This is called dispersion. The amount the prism or grating spreads the light depends on many factors that will be explained later. Dispersion, or plate scale (not to be confused with resolution), is usually measured either in Å/mm or Å/pixel (nm/mm or nm/pixel) (when using an imaging camera). A low dispersion system would be 200 Å/mm (20 nm/mm) and high dispersion < 10 Å/mm (1 nm/mm). As the amount of available light, say from a star, is limited, and not all this light ends up in the spectrum; the lower the dispersion the brighter the spectrum appears. Prisms unfortunately have a non-uniform dispersion where the blue light is dispersed more than the red; this can make subsequent analyzing and measuring wavelengths of the spectrum a bit more difficult.

K.M. Harrison, *Astronomical Spectroscopy for Amateurs*, Patrick Moore's Practical
Astronomy Series, DOI 10.1007/978-1-4419-7239-2_4,
© Springer Science+Business Media, LLC 2011

Depending on the size of the CCD sensor more or less of the spectrum will be recorded on the image. With larger dispersions the grating can either be set in a position to record a specific wavelength (i.e., Hα) or mounted on a turntable to allow it to be rotated and bring various sections of the spectrum onto the CCD (For example, a MX7c-sized chip and a plate scale of 10 Å/mm (1 nm/mm) requires some twelve images to cover the full extent of the visible spectrum).

Like telescopes, the measure of the detail seen in the spectrum is limited by its resolving power. In the case of a spectrum this is the ability to detect differences between very close features in the spectrum, such as absorption bands. A good test for small spectroscopes is the ability to resolve the sodium double at 589.0 and 589.6 nm.

The theoretical spectral resolution (R) is defined as:

$$R = \lambda/\Delta\lambda$$

where $\Delta\lambda$ is the minimum distance between two features in the spectrum which are resolved. A low resolution spectroscope would have an $R = 100$ (resolving 6 nm at 600.0 nm), whereas a high resolution one can be at $R = 10000–20000$ (0.06–0.03 nm at 600.0 nm). To get a high resolution generally means using a high dispersion system.

Sometimes you'll see resolution stated in Å/pixel. This obviously depends on the size of the CCD camera pixel and the "Nyquist sampling" requirement, where the illumination of at least two pixels is need to accurately record the detail. Full width half maximum (FWHM) of a particular line feature is also used as a measure of resolving power.

The spectral resolution can be influenced by factors such as:

- size of the star image (for transmission gratings)
- distance from the grating to the camera/CCD (for transmission gratings)
- optical aberration in the system
- number of lines on the grating
- the width of the slit (when used)
- the focal length of the collimator/camera lens
- the pixel size on the CCD

These factors will be discussed in detail when considering the specific design of each type of spectroscope in Part III of this book.

Efficiency of Your Spectroscope

Not all the available starlight arrives in the final spectrum, due to light loss through the spectroscope. The overall efficiency (sometimes called "throughput," "light-gathering power," or "etendue") varies due to factors such as:

- entrance slit size
- reflections and light loss through the collimating or camera lens
- vignetting of either the collimating or camera lens
- vignetting of the grating
- reflection efficiency of the grating
- order of spectrum being observed
- effective "blazing"
- quantum efficiency of the CCD sensor
- losses in fiber optic connections (if used)

These factors can quickly reduce the overall optical efficiency to very low numbers, and 10–15% overall transmission of the available starlight to the final CCD image is not uncommon. These factors will be discussed in more detail in Part III of this book.

Prisms

Prisms have been used since the time of Newton to disperse white light into a spectrum. The dispersion of a prism results from it having a different refractive index for each wavelength, and the total dispersion depends on the type of glass and the angles of the faces. By balancing the refractive index (n) of different glasses, astronomers can design color-correcting objectives – but in our case we want the opposite! We want to *generate* color.

The relationship between the refractive index and the deviation of the light beam was first determined in 1621 by Willebrord Snellius (1580–1626) and is measured using Snells Law:

$$n = \sin i / \sin r$$

The *i* is the angle of incidence to the surface.
The r is the angle of refraction from the surface.

This refraction occurs at both the entry surface and exit surface of the prism. For minimal deviation and optimum image quality the light passing through the prism should be parallel to the bottom surface. The incident and exit angles will be approximately half the prism apex angle (see Fig. 4.1).

In 1896 J. F. Hartmann (1865–1936) investigated the dispersion of a prism and developed his dispersion formula, which provides an empirical solution to the change in refractive index with wavelength. In a dense flint prism, the deviation of blue light is almost five times that of red light.

Most of the prisms readily available to the amateur are BK7 glass. This is a clear crown glass that transmits light (more than 98%) from around 390 nm through to the infrared.

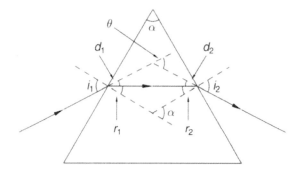

Figure 4.1. Path of a monochromatic light beam through a prism.

Any prism can be used to make a simple spectroscope – for example 45° prisms from old binoculars. Better still is a flint F2 or SF2 60° prism.

Resolution

The resolution of the prism spectrum is generally dependent on the star image size or the slit width (if used), the base length of the prism, the focal length of the camera lens, and the pixel size of the CCD.

To improve the spectral resolution from a prism, multiple prisms can be used to increase the effective base length, each one set at the minimum deviation angle to the previous. Early spectroscopes used up to six or even eight prisms in the "train" to get maximum resolution.

The theoretical resolution of a prism can also be stated as:

$$R = \Delta n / \Delta \lambda * \text{base length of prism(s)}$$

where $\Delta n / \Delta \lambda$ is the dispersion of the prism.

Efficiency of a Prism Spectroscope

One of the strengths of a prism-based spectroscope is the fact that only one spectrum is produced, unlike the multitude of spectra from a grating. For single prism system this gives a brighter spectrum. Larger prisms and the use of multiple prisms reduce this benefit due to the additional light losses both from the surfaces and the thickness of the glass.

Amici Prism Systems

An Amici prism is made up from three small prisms in a crown/flint/crown combination. This gives zero deviation and a dispersion similar to a small crown prism of half its length. A theoretical resolution of $R = 400$ (1.3 nm) is not uncommon. Most of the direct vision spectroscopes available today use Amici prisms. See Fig. 4.2.

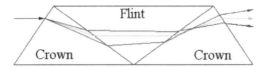

Figure 4.2. Amici prism. (I. Galidakis.)

Gratings

Gratings are widely used as an alternative to prisms to achieve an almost linear dispersion. See Fig. 4.3.

A word of caution – all gratings are very delicate and should be handled with extreme care at all times. NEVER touch the grating surface! There is really no satisfactory method of cleaning a grating. Minor dust particles can be blown off the surface (use a proprietary blower, not the mouth!) but other than that, leave them well alone and keep them in a dust-free sealed container at all times when not in use.

Fraunhofer's earliest gratings were made by winding very fine wire around a two-part frame. When the sections were separated, he obtained a transmission grating with approximately 150 lines per mm (l/mm). It was with such a grating that he measured the positions of the absorption lines in the solar spectrum.

Rowland developed a "ruling engine," basically a fine-shaped diamond point held in a traveling arm that could be accurately moved back and forth with a screw, providing a small index movement between strokes. This then generated a lined surface, initially on a glass plate and later on aluminum or magnesium plates. Since the diamond quickly wears out, the early gratings were small and inefficient. By the end of the nineteenth century, however, it was possible to produce accurate gratings of up to 6″ wide with lines at 1500 l/mm.

This manufacturing technique is still used today to produce "master gratings." These are then coated with a plastic solution that faithfully replicates the grating. This film can then be mounted on glass as a transmission grating or aluminized to form a reflection grating.

Gratings in which the grooves scatter the incident light (to cause interference patterns) are said to be "amplitude" gratings. The majority of the gratings used by amateurs are of the "amplitude" type.

Figure 4.3. Selection of transmission gratings. (Courtesy of Paton Hawksley.)

A recent alternative is to use the interference pattern between two laser beam to "photo-etch" a substrate and give a sinusoidal grating shape. These holograph gratings are very accurate and can be made in large sizes. Generally all gratings above 1800 l/mm are now holographic gratings.

New technology holographic gratings that use refractive index modulations within a thin layer of material sandwiched between two glass substrates to change the diffraction of the incident light are known as "volume-phase" (VP) gratings. The intensity of the refractive index modulation and the depth of the grating layer are critical design factors.

Grating Theory

When a collimated monochromatic light beam goes through a pair of fine slits, the action of the light waves causes both diffraction and interference effects. As the light waves exit the slits they interact at varying angles from the slit to generate a series of interference patterns, where the crest of one wave reinforces the other, giving a bright line or no light where the crests oppose each other. When white light is considered and there are a large number of slits (i.e., grooves in the grating) these

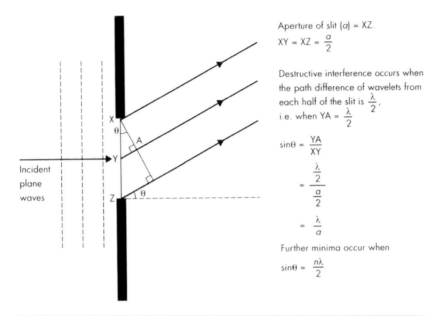

Aperture of slit $(a) = XZ$.

$$XY = XZ = \frac{a}{2}$$

Destructive interference occurs when the path difference of wavelets from each half of the slit is $\frac{\lambda}{2}$, i.e. when $YA = \frac{\lambda}{2}$

$$\sin\theta = \frac{YA}{XY}$$

$$= \frac{\frac{\lambda}{2}}{\frac{a}{2}}$$

$$= \frac{\lambda}{a}$$

Further minima occur when

$$\sin\theta = \frac{n\lambda}{2}$$

Figure 4.4. Grating – path length difference. (S. Tonkin.)

interference patterns vary with wavelength and give rise to a series of spectra. A grating can produce up to 10 (sometimes many more) orders of spectra. See Fig. 4.4.

For a transmission grating, some of the light source is un-deviated as it passes through the grating and gives a "zero order" image. Spectra are formed symmetrically around this zero order image at various angular distances, the 1st order being the brightest.

The grating equation is as follows:

$$nN\lambda = \sin\alpha + \sin\beta$$

This defines the relationship between the angle of the incident beam (α), the angle of the exit (or reflected beam) (β), the spectrum order (n), the grating l/mm (N), and the wavelength of the light (λ).

Due to the linear dispersion of gratings, the 2nd order spectrum is twice the length of the 1st order and can overlap on the red portion of the 1st order (The 400 nm in the 2nd order will be produced at the same point as the 800 nm in the 1st order). The non-overlapped length of the spectrum defines the free spectral range of the grating. See Fig. 4.5.

The interference in the red portion of the 1st order spectrum by the blue region of the 2nd order spectrum can be easily suppressed by using a red filter (A Wratten 25A filter will allow imaging of the Hα (656.3 nm) region without 2nd order interference).

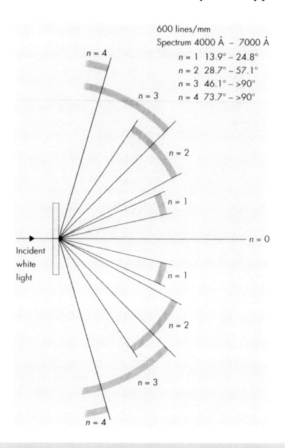

Figure 4.5. Showing spectra spread from a 600 l/mm grating. (S. Tonkin.)

For a grating positioned square to the incoming light ($\alpha = 0$), the spectral deviation of the 1st order spectrum ($n = 1$), a 100 l/mm grating (N) based on 550 nm (λ) is:

$$Sin\beta = nN\lambda$$

$$Sin\beta = 1^*100^*550^*10^{-6}$$

$$\beta = 3.15^\circ$$

Transmission gratings with low l/mm have a small deviation angle that allows them to be used in line with the camera with minimal distortion. Above 200 l/mm the camera should be set at the deviation angle to the grating to give best results.

For green light (550 nm) the deviation for various gratings is:

100 l/mm	3.15°
200 l/mm	6.3°
400 l/mm	12.7°
600 l/mm	19.3°

The spectral dispersion is:

$$d\lambda/d\beta = Ncos\beta/n$$

For small angles, cosβ tends to 1, giving the dispersion as:

$$d\lambda/d\beta = N/n$$

$$\text{Linear dispersion or plate scale} = 10^7cos\beta/n \, N \, L \, \text{Å/mm}$$

where, L, is the distance from the grating to the CCD.

For a 100 l/mm grating and 1st order spectrum, at a distance of 80 mm, this gives: plate scale = 10^7 cos(3.15°)/1 * 100 * 80 Å/mm = 1248 Å/mm

The dispersion of the 2nd order is twice that of the first, so a 600 l/mm grating working in the 2nd order will have the same dispersion as a 1200 l/mm working at 1st order.

The resolution based on the Rayleigh criteria is:

$$\lambda/\Delta\lambda = nN = R$$

More illuminated lines on the grating gives more resolution; hence use the largest grating you can to get maximum theoretical resolution.

The light distribution across the zero image and subsequent spectra is approximately 50% in the zero image, 15% in each of the 1st order, 5% in the 2nd, and so on. This is very inefficient, and to improve this distribution most professional gratings are now "blazed". Blazing is where the shape of the groove in the grating is tilted to preferentially deviate the maximum amount of light into one of the 1st order spectra. This can dramatically improve the efficiency, and up to 70% of the incoming light can be directed into the 1st order.

Gratings are available in various l/mm, typically 100, 200, 300, 600, 1200, 1800, and 2400 l/mm. The dispersion for a 200 l/mm grating will be twice that of the 100 l/mm, so the spectrum formed will be fainter but the resolution higher.

Standard sizes for gratings are 25 mm×25 mm, 25 mm×30 mm, 25 mm×50 mm, 30 mm×30 mm, and 50 mm×50 mm. The price increases exponentially with size, and most amateurs use either the 25 mm×25 mm or the 30 mm×30 mm gratings. Transmission gratings of 100 l/mm and 200 l/mm are available mounted in standard 1.25″ filter size (28.5 mm thread) (i.e., Star Analyser, Rainbow Optics, Baader, etc.).

Depending on the area of interest and the resolution required, different grating l/mm can be used. For instance, a 100 l/mm grating will allow the amateur to see the main absorption lines in the brighter stars and confirm their spectral classification, but for analysis of the profile of the Hα line for Be stars, etc., a 1200 l/mm or higher grating is required.

Although it is said that the grating dispersion is linear, this is not strictly true. The spectrum is formed at the deviation angle, and unless the camera is correctly readjusted between the blue and red, some of the spectrum will be out of focus. Focusing on the zero order image will *not* give a good focus on the spectrum.

Gratings also distort the shape of the light beam (anamorphic factor), and this needs to be considered in the final design of a spectroscope. Detailed discussions are given later in this book.

Further Reading

Hearnshaw, J. B. *Astronomical Spectrographs and Their History.* Cambridge University Press (2009).

Web Pages

http://ioannis.virtualcomposer2000.com/spectroscope/amici.html#diagram
http://www.noao.edu/noao/noaonews/jun98/node4.html

CHAPTER FIVE

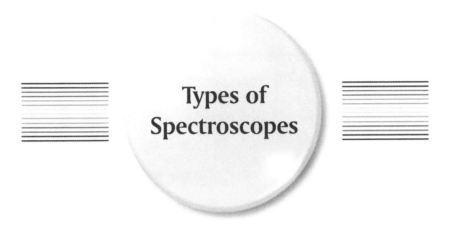

Types of Spectroscopes

There are basically four or five designs of spectroscope of interest to the amateur astronomer. Each has its use, and each varies greatly in complexity, spectral resolution, and type of object that can be satisfactorily observed.

Objective Prism or Grating

When a prism or transmission grating is placed directly in front of the telescope objective or camera lens, it becomes an objective prism/grating system. See Fig. 5.1. This was the earliest use of the spectroscope and was successfully used to catalog the stellar classifications. Small angle prisms the same size as the telescope objective are ideal. When this can't be achieved, smaller prisms can be mounted in a cover frame in front of the telescope or camera lens. Good results have been achieved using readily available 30 or 45° prisms with 135–200 mm telephoto lenses.

Alternatively a transmission grating can be used. Generally the lower l/mm gratings give better results (i.e. <300 l/mm). The size of the spectrum and the plate scale produced will depend on the size of the prism/grating and the focal length of the telescope. A 100 l/mm grating at a distance of 100 mm will give approximately the same plate scale as a 30 mm 60° flint prism.

Overlapping spectra and star images in spectra can cause difficulty in subsequent analysis, but by re-positioning the grating line angle relative to the star images, this can be minimized.

K.M. Harrison, *Astronomical Spectroscopy for Amateurs*, Patrick Moore's Practical Astronomy Series, DOI 10.1007/978-1-4419-7239-2_5,
© Springer Science+Business Media, LLC 2011

The field of view (and the star images) will be off-set to the optical axis of the telescope by the deviation of the prism or grating (approximately half the prism angle, i.e., 15° for a 30° prism); this needs to be taken into account when setting up.

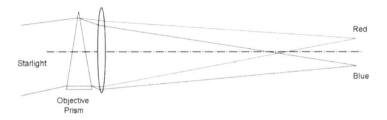

Figure 5.1. Objective prism layout.

Prism with Collimator/Camera-CCD

A prism can be mounted in the telescope behind a Barlow lens or a positive lens where the output is a collimated beam. An imaging lens positioned at the deviation angle followed by a CCD camera will provide low to medium spectral resolution images. In this position the prism acts like a "virtual" objective prism. See Fig. 5.2.

Similarly, an Amici prism (taken from a D-V spectroscope) can be used.

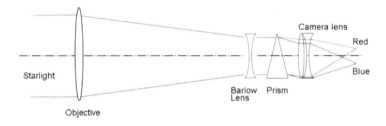

Figure 5.2. Prism in collimated beam.

Traditional Prism Littrow

A 30° prism with a silvered rear surface was originally used in the Littrow spectroscope design, developed by Otto Littrow (1843–1864) in 1863. Figure 5.3 shows the optical layout. An entrance slit (S) is positioned at the focus of the telescope, and the incoming light is deflected by a small mirror or prism (R) positioned just above

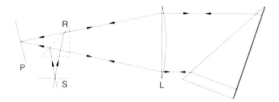

Figure 5.3. Original Littrow spectroscope design. (*Spectrophysics*, Springer, 1989.)

the optical axis, through a collimating lens (L) to the prism. The reflected light, now dispersed, travels back through the collimator lens to the imaging camera (P).

The rear reflection gave the equivalent dispersion of a 60° prism, with the advantage that the collimating lens also acted as the camera lens, resulting in a robust, compact arrangement. Nowadays a similar optical arrangement is used but the device commonly has a reflection grating instead of the prism (i.e. Shelyak LhiresIII).

Transmission Grating in Converging Beam

The popularity of the 1.25″ filter-sized transmission gratings as a "first spectroscope" has been growing over the past few years and has provided a good entry point for budding amateurs interested in astronomical spectroscopy. These gratings are well suited to stellar spectral imaging, but being a "slit-less" design, they are of limited use on extended objects such as gaseous DSO's. These gratings are usually 100 or 200 l/mm and are used on the telescope in the converging beam. See Fig. 5.4.

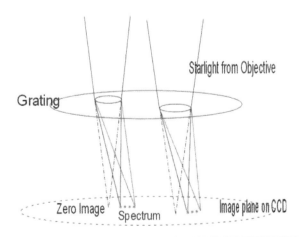

Figure 5.4. Transmission grating in a converging beam.

When positioned in front of a low power eyepiece they can give pleasing visual results. A 100 mm telescope will show the absorption lines of the brighter stars. They can also be mounted in front of a camera/CCD fitted with a lens to image stellar spectra. The best results are obtained with smaller f/ratios, which give smaller star images. For any particular grating, the linear dispersion/plate scale is proportional to the distance from the grating to the film plane.

We know from earlier, with 100 l/mm grating, used in a 200 mm SCT at f6.3 and a grating distance of 80 mm with a DSLR (7.4 μm pixel), the plate scale is 1,260 Å/mm, or 9.23 Å/pixel. Grating distances of 30–80 mm are regularly used. If the CCD sensor is large enough, it's possible to record the star's zero image and its spectrum at the same time. This makes subsequent analysis easier.

The resolution is restricted by the size of star image, telescope optical aberrations, distance to CCD, and the number of lines illuminated. The size of the star image effectively becomes the entrance slit, and obviously the smaller the image the better the resolution. Depending on the system a maximum of $R = 100$ to $R = 200$ (i.e., a resolution of 60–30 Å) can be achieved. There has been some success with the use of an aperture mask (a disk with a small hole, approximately 6 mm diameter, positioned off-axis) to restrict the field of view and keep interference from other secondary stars and spectra to a minimum.

Transmission Grating in Collimated Beam (with or without Slit)

If a collimating lens is added in front of a transmission grating (either a positive or negative [Barlow] lens) and a camera lens/CCD mounted behind the grating, then the resolution can be increased. This is similar to the arrangement shown in Fig. 5.3 but with a grating replacing the prism.

The size of the star image will limit the resolution (typically 2–4 arc sec) unless a slit is used at the focus of the telescope. Adding a slit compounds the problems of finding the star and guiding but can improve the resolution. Again, with higher l/mm gratings the camera must be positioned at an angle to the optical axis (or some means of rotating the grating included in the design).

Grisms

To reduce the camera/CCD alignment problems associated with the deviation angle from a transmission grating (and reduce the effects of chromatic coma in a converging beam layout), a small-angle prism can be mounted just in front of the grating to compensate and apply a reverse deviation, bringing the image of the spectrum back onto the optical axis. This configuration is called a "grism" (grating prism). The deviation angle of the prism should be approximately

$$\text{Prism deviation angle} = 3.8° \times \text{l/mm}/100$$

This gives a more uniform spectrum on the CCD. See Fig. 5.5. Calibration and analysis is then done using either reference lamps (if a slit is fitted) or recognized spectral features, i.e., Balmer series. The introduction of the prism leads to a slightly non-linear spectrum. This procedure is generally used on >200 l/mm gratings.

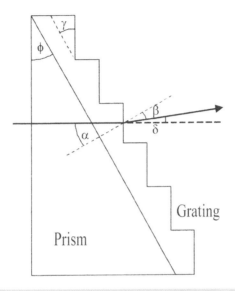

Figure 5.5. Grism optical layout. (J. Allington-Smith.)

Reflection Grating (with or without Slit)

Reflection gratings can be used in various configurations, i.e., classical, Littrow, Ebert-Fastie, Czerny-Turner, etc. Most amateur-sized instruments use plane gratings rather than curved gratings. Generally they have entrance slits, a collimating lens, and a camera lens to record the spectrum. Mirrors can be used instead of lenses.

The spectroscope is mounted on the telescope with the entrance slit positioned at the prime focus and the star image focused on the slit.

This is a basic spectroscope design where a chromatic doublet is used as a collimator, reflection grating, and a further doublet as the imaging lens. See Fig. 5.6. A popular variation on this design is to replace the collimator lens with a spherical mirror.

The Littrow design is widely used with a grating replacing the original prism. A spherical mirror can be used as a collimator to reduce chromatic aberrations, as shown in Fig. 5.7.

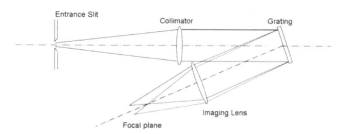

Figure 5.6. Classical spectroscope, optical layout.

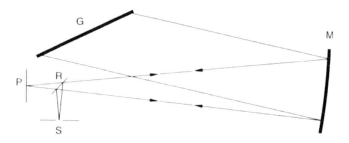

Figure 5.7. Littrow optical layout. (*Spectrophysics*, Springer, 1999.)

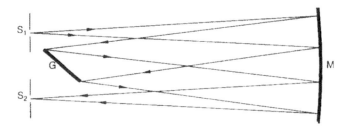

Figure 5.8. Ebert-Fastie optical layout. (*Spectrophysics*, Springer, 1999.)

The Ebert-Fastie (after Hermann Ebert, 1861–1913, and William G. Fastie, 1916–2000) uses a single large spherical mirror to act as a collimator and to focus the image on the spectrum. See Fig. 5.8. Light enters through the slit (S_1) positioned at the focus of the mirror (M) and is reflected as a collimated beam to the grating (G); the dispersed light is then focused by the mirror, forming an image of the spectrum at S_2.

The Czerny-Turner (after Marianus Czerny, 1896–1985) and Arthur F. Turner, (1906-1996), who developed the design in the 1930s) is an improvement on the earlier Ebert-Fastie design and utilizes two smaller spherical mirrors (M_1, M_2) to reduce aberrations. See Fig. 5.9.

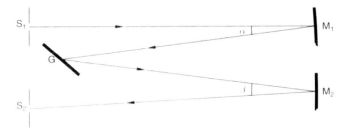

Figure 5.9. Czerny-Turner optical layout. (*Spectrophysics*, Springer, 1999.)

When used without an entrance slit the resolution is limited by the stellar image size at the telescope focus. Adjustable slits can allow the spectroscope to be used in a wide slit configuration (i.e., the slit gap is much wider than the star image), which can help to reduce the background noise in the spectrum.

Reflection slits, where the front surface has a mirror finish, can be used to provide a guide image of the star under observation. Unfortunately these are not easily obtained by the amateur constructor, tend to have limited range of aperture sizes (i.e., 10, 20, or 30 μm slit width), and be very expensive.

The collimating lens/mirror must be matched to the telescope focal ratio to minimize transmission losses and vignetting. If the spectroscope is used, say, on an f6 telescope, then the collimator should also be f6.

The grating needs to be large enough to accept the collimated beam without vignetting and mounted in a support frame that does not subject it to unnecessary forces. Where imaging of the whole spectral range is required, the grating support needs to be able to rotate to bring different parts of the spectrum onto the CCD. For maximum efficiency, blazed gratings used in the 1st order spectra are generally preferred. Depending on the focal lengths of the collimating/camera lens some instruments can be quite large and pose mounting and balancing problems on the telescope. Amateur-sized spectroscopes (200 mm collimator focal length) can weigh up to 3 or 4 kg.

With larger spectroscopes, optical fiber couplers are sometimes used to link the telescope to the instrument, which is then located remotely (but nearby). The use of optical fibers is discussed in more detail later in the book.

Echelle (from the French for ladder) spectroscopes use a very coarse (30–80 l/mm) high-efficiency blazed grating to give high dispersions over a short wavelength range. When used in the high spectral orders (n>40) the free spectral range is very low (100 Å), and the overlapping spectra are separated using a second grating (or prism) at right angles. The final spectrum produced is a matrix of short spectral sections and can give very high resolution (R>50000). See Fig. 5.10.

The extremely high blaze angle of the Echelle grating concentrates the energy in the higher orders. See Fig. 5.11.

From the spectral deviation equation we saw earlier, it follows that in higher orders the angular separation between two wavelengths becomes greater. Consider

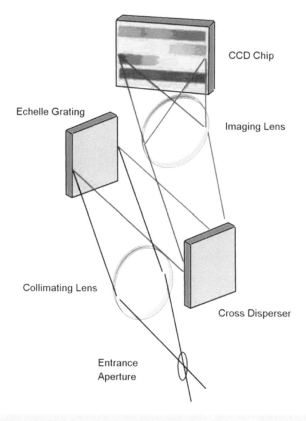

Figure 5.10. Echelle spectroscope optical layout.

Figure 5.11. Echelle spectrum matrix. (M. Dubs.)

two lines, one at 600 nm and the other at 605 nm, incident on an Echelle grating with 79 l/mm. At $n = 1$ the angular separation is 0.03°, but at $n = 40$ the angular separation becomes 2.1°. The disadvantage is the reduced free spectral range, which decreased from 630 nm (630 nm/1) to only 15.6 nm (630 nm/40).

Further Reading

Hearnshaw, J. B. *The Analysis of Starlight.* Cambridge (1986).
Kitchin, C. R. *Optical Astronomical Spectroscopy.* Taylor & Francis (1995).
Kitchin, C. R. *Astrophysical Techniques.* Taylor & Francis (2003).
Thackery, A. D. *Astronomical Spectroscopy.* Eyre & Spottiswoode Ltd (1961).

Web pages

http://www.cfai.dur.ac.uk/old/projects/dispersion/grating_spectroscopy_theory.pdf
http://bass2000.obspm.fr/solar_spect.php?step=1
http://www.horiba.com/scientific/products/optics-tutorial/diffraction-gratings/
http://refractiveindex.info/index.php?group=SCHOTT&material=N-SF2&wavelength=0.4
http://www.regulusastro.com/regulus/spectra/index.html
http://zebu.uoregon.edu/spectra.html
http://nedwww.ipac.caltech.edu/level5/ASS_Atlas/frames.html
http://outreach.atnf.csiro.au/education/senior/astrophysics/spectra_astro_types.html
http://laserstars.org/spectra/index.html
http://fermi.jhuapl.edu/liege/s00_0000.html/
http://spectroscopy.wordpress.com/instruments/
http://www.peripatus.gen.nz/Astronomy/SteCla.html
http://las.perkinelmer.com/content/TechnicalInfo/TCH_Raman400Echelle
    Spectrographs.pdf

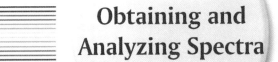

PART TWO

Obtaining and Analyzing Spectra

CHAPTER SIX

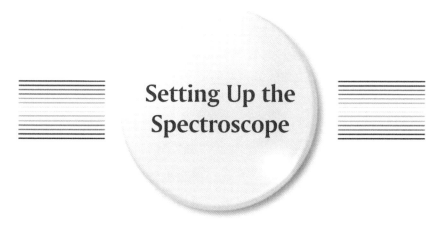

Setting Up the Spectroscope

Objective Prisms/Gratings

Getting and recording images of spectra is what it's all about. The easiest way to start is to use your prism or transmission grating in front of the camera lens or telescope objective. This objective prism/grating setup can allow you to record the stellar absorption and emission lines at low resolution. You'll need a tracking platform or equatorial mounting properly polar aligned.

Camera Lenses

Fixed focal length lenses are preferred to zoom lens. Any lens from 50 mm focal length upwards can be used. Set to widest aperture.

Telescopes

Almost any telescope can be used with an objective prism/grating. With Newtonians, Maksutov's, and SCT's, the prism should be mounted off center to avoid interference from the secondary mirror. The longer focal length will give the spectrum a larger image scale, and coverage of the whole spectrum needs to be checked against the size of the CCD chip.

K.M. Harrison, *Astronomical Spectroscopy for Amateurs*, Patrick Moore's Practical Astronomy Series, DOI 10.1007/978-1-4419-7239-2_6,
© Springer Science+Business Media, LLC 2011

Mounting the Objective Prism/Grating

Objective Prisms

When using camera lenses, the prism can be mounted in an old lens hood or a special holding frame fabricated from thin plywood/MDF. Ideally the prism should be tilted by half its deviation angle to keep the spectrum image close to the optical axis.

For refracting telescopes the small aperture cover usually found in the main dust cover can be used (or "modified" in the case of Newtonians, etc., by cutting a suitable-sized hole off center). The prism can be held securely in place with Gaffa tape or small clamps. Alternatively a lightweight mask can be made up (such as of thin ply) to hold the prism in place. When in final position, the prism should be orientated along the Dec axis; this allows any inconsistencies in the RA drive to widen the spectrum rather than blur the image. Likewise changing the RA drive rate slightly will have the same effect. Make sure the prism is firmly secure before you finish. See Fig. 6.1.

Figure 6.1. Mounting an objective prism.

Objective Gratings

With the smaller transmission gratings (Star Analyser/Rainbow Optics) (see Fig. 6.2), these can be mounted onto the camera lens by cutting a 28.5 mm-diameter hole in the center of an old lens cap and "screwing" the grating into the hole. An

Figure 6.2. Filter gratings.

alternative is to mount the grating in a 1.25″ T thread nosepiece and use a T2 to 48 mm (2″ astronomical filter thread) together with any other stepping rings required to fit the front of the lens.

You can also separate the inner T thread section from a T2 adaptor; unscrew the three small grub screws on the outer rim, and the internal T thread section will fall out. The outside diameter of this insert is almost exactly 49 mm. This is ideal for gluing into the center of a 52–55 mm step up ring. Additional stepping rings can then be used to bring this adaptor to the final camera lens size. See Fig. 6.3.

HINT: If you try this method, do a "dry fit" first, i.e., grating mounted into the 1.25″ T thread adaptor plus the T thread insert – setting in the step up ring plus any additional stepper rings. Mount this carefully onto the camera lens and mark the position of the T thread insert and the 52 mm adaptor where the grating is dispersing horizontally across the CCD chip. When gluing, use an epoxy-type glue, and line up the marks. This will make it much easier to use the grating at night without worrying about its alignment to the camera.

A similar arrangement to that detailed for prisms can be used to hold the grating in front of the telescope objective. Leaving the grating in a 1.25″ nosepiece is a safe option and allows the T section base to be used for mounting.

Using small prisms and transmission gratings obviously reduces the effective aperture and light-gathering capabilities of the camera/telescope, but can still produce some excellent results on the brighter objects.

Larger prisms and gratings can be mounted in the same fashion as the smaller ones. Watch for balance problems with large prisms; they can be heavy!

Figure 6.3. Filter grating mounted on camera lens.

Adjustments and Focus

Prisms

The direction of the starlight recorded will be at the deviation angle of the prism; this means that the object being observed will appear at some angle from the optical axis and alignment of the camera body. For a 30° prism this angle is approximately 15°, so, to align with the tracking telescope the camera needs to be moved by this amount. As mentioned earlier, it's better if the dispersion is aligned along the RA axis, so this offset-angle would also be in the RA direction.

Rotate the camera body to bring the spectrum horizontal to the CCD, without upsetting the alignment of the prism to the sky. Make sure the camera is firmly secure and doesn't tend to sag or rotate on the camera support. See Fig. 6.4.

When you use prisms, there's no zero order image to align on (or to focus on!), only – hopefully – a bright spectral band to guide you. Set the lens on infinity focus and take a few test images to check the results. When you find the best focus position tape the lens focusing ring with a piece of masking tape to make sure it doesn't move.

Setting the camera to ISO200, Bulb setting and maximum aperture exposures of a few seconds will record spectra. A remote shutter trigger is very useful to reduce vibration.

For CCD (or webcam) cameras, start with a 1-sec exposure on a bright star; the spectrum should be clear and not overexposed.

Figure 6.4. Objective prism mounted on a piggyback camera. Note the off-set angle.

Gratings

Using a grating allows us to see the FOV zero image stars. This makes locating targets and initial focusing a bit easier. When imaging, the deviation angle of the grating needs to be considered. Will the camera/CCD chip be large enough to record both the zero image and the spectrum? See Figs. 6.5 and 6.6.

For short focal length objectives and a DSLR camera this is not usually a problem, but with smaller CCD chips you need to do a quick calculation. For green light (550 nm) for the various gratings here is what you need to know:

100 l/mm	3.15°
200 l/mm	6.3°
400 l/mm	12.7°
600 l/mm	19.3°

Figure 6.5. Cokin Cosmos grating filter mounted in lens hood of a 135 mm telelens.

Figure 6.6. P-H grating mounted on a 135 mm telelens.

Also, due to the deviation and dispersion angles (field curvature) the blue focus will vary from the red focus. For general low resolution full width spectra, focus around the yellow/green as a compromise. The zero image may appear out of focus, but this is OK.

A few trial exposures will quickly guide you to the best focus position. We'll discuss processing of the spectra later.

How do you know which object is in the field? The field of view through the prism/grating will be similar to the camera/telescope FOV with an eyepiece that has a focal length that is double the CCD chip width. For example, a webcam with a 3 mm chip visually shows the same FOV as with a 6 mm eyepiece. The appropriate FOV can be used to make overlays for a star atlas or programmed into the planetarium package you use to confirm the target objects. *Uranometria* star charts and Carte du Ciel software are recommended (Finding target objects, especially very faint stars, becomes even more important when using larger slit type spectroscopes!!).

HINT: When using a piggyback camera arrangement on a GOTO mounting, the camera deviation offset can be "corrected" by setting the telescope on the center of the field to be imaged (or a bright star nearby), moving the axes manually to the offset angle (check that the anticipated spectra are visible!), then re-setting the GOTO by synchronizing the original position on the new settings. This will give you a readout at the telescope and planetarium software of the actual stars/FOV being recorded. You can then slew to the next object/field using the GOTO.

Flats and Darks

As in conventional astronomical imaging the use of flats and darks can improve the quality of the spectrum recorded. Flats can be taken using a "flat spectrum" halogen lamp. These should be re-done for any subsequent change in the optical configuration. See later chapters for more detail.

Web Pages

http://www.starrywonders.com/ccdcameraconsiderations.html

CHAPTER SEVEN

Using Spectroscopes in the Converging Beam

Before considering the larger reflection grating spectroscopes, let's look at a popular method of using the smaller transmission gratings – i.e., Star Analyser 100, Rainbow Optics 200, Baader 200, and Spectral Optics – in the converging light beam.

All of these gratings are designed to fit a standard 1.25″ filter thread and can be used visually with a low power eyepiece to show stellar spectra. All allow the observer to easily distinguish small planetary nebulae (emissions in discrete wavelengths). A small cylindrical lens held between the eye and the eyepiece will "stretch" the spectrum into a broader band, making any lines easier to see (Such lenses are supplied with the Baader and Rainbow optics gratings).

They can also be fitted into a 1.25″ T thread adaptor, which is then mounted to the CCD camera (or via a T ring to the DSLR camera body) This positions the grating approx 50–80 mm in front of the CCD chip. Alternatively the grating can be fitted into a standard 1.25″ filter wheel in front of the camera body. Either way, the grating is positioned in the converging beam between the telescope objective (or mirror) and the CCD chip at the prime focus. See Figs. 7.1 and 7.2.

As mentioned earlier, the faster the f/ratio the smaller the star image. On SCT's it's therefore beneficial to mount the grating behind a focal reducer (i.e., an × 0.63 reducer). The grating produces a "zero" order image of the star being observed and the bright 1st order spectrum lies off center (at the deviation angle) from the optical axis of the telescope. The deviation angle, β, can be calculated from an equation given earlier.

The final spectral resolution of this arrangement is limited by the seeing conditions, optical aberrations, and pixel size of the CCD. The actual resolutions will vary from 18 to 40 Å, depending on the grating, etc.

K.M. Harrison, *Astronomical Spectroscopy for Amateurs*, Patrick Moore's Practical
Astronomy Series, DOI 10.1007/978-1-4419-7239-2_7,
© Springer Science+Business Media, LLC 2011

Seeing	Theoretical resolution (Å)
2″	18
3″	18
4″	28
5″	37

Figure 7.1. Filter gratings fitted to a QHY5 camera and a Philips SPC900 webcam.

Figure 7.2. Filter grating fitted to a Canon 300D DSLR.

Obtaining Your First Spectrum

Set up the telescope as usual and point it towards a bright star above say, a 45° elevation (This reduces the effects of the atmospheric absorption and should give better seeing conditions).

Without the grating attached, insert the camera/CCD with its nosepiece and focus on the star. Use your normal method, i.e., Bahtinov mask, inspecting the image on the PC screen to find focus. The final position of the focus will be shifted outwards slightly when the grating is attached, so don't worry too much about getting an accurate focus at this stage.

Remove the camera/CCD without touching the focus settings. Insert the grating and rotate it to get the grating lines vertical to the image frame. If you slightly unscrew the filter grating from the nosepiece and wrap a couple of turns of thread onto the grating threads, these can help to give some "fine adjustment" to the grating orientation. Re-insert the camera/CCD. Looking at the CCD image on the PC (viewfinder on a DSLR) you should see a star image close to the center of the frame and spectra at either side. If you are using a blazed grating, one spectrum will be brighter than the other. Using the telescopes slow motion controls, bring this brighter spectrum close to the middle of the frame. The star (zero image) will be off to the side. Keep moving the star image until it is just inside the frame.

Now study the star image and the spectrum. As a minimum the star image should now be brought to a precise focus. Better still is to adjust the focus until the spectrum band is as narrow as possible, with clear sharp edges. Check this focus using an initial exposure of, say, 1 sec, and that the spectrum is lying horizontally across the frame. If you find that secondary star images/spectra are interfering with the target spectrum, try rotating the camera body (without changing the grating to CCD orientation) in the focuser to move the background stars away from the target spectrum.

Take a couple of exposures, checking if the spectrum is over- or underexposed. On very bright stars (i.e., Vega) you may find you only need a 0.2- or 0.4 sec exposure.

Now that you have a spectral image, open it in your usual imaging software, zoom in on the spectrum, and see if any dark/bright lines or bumps can be seen. Compare the exposures looking for best focus and clarity.

On the brighter A-type stars the dark hydrogen Balmer lines should be visible. Note the setting of your focuser and the best exposure times for future reference.

Points to Note

If the spectrum is recorded on your CCD chip at an angle you can lose resolution when rotating it horizontally. Consider that 5 or 6 vertically inclined pixels will record a particular line when the "actual" width was only 2 pixels. When these are rotated into the horizontal the line will be "spread" over these pixels, reducing contrast and resolution.

You can use the FWHM criteria to establish the best focus on the zero image. Most image capture and processing software (AstroArt, Maxim DL, etc.) can give this FWHM information. Obviously the lower the FWHM value, the better the focus.

The distance from the grating to the CCD chip can be varied by either adding 1.25″ filter spacer rings or T thread spacers behind the nosepiece. Try various settings, keeping the zero image in the frame if possible. Distances above 80 mm tend to be counterproductive, as the benefits of the larger dispersion are outweighed by the inherent optical aberrations.

The effects of field curvature and chromatic coma can significantly reduce the resolution. These effects can be minimized by repositioning the CCD plane for field curvature and adding a small wedge prism/grism for chromatic coma.

Like astrophotographic images, short spectral exposures (taken with exactly the same grating settings) can be stacked to improve the signal/noise and allow the spectra of fainter objects to be imaged.

To get a good spectrum of, say, a 7 mag star with a 100 l/mm grating you may need a total exposure of up to 5 min.

If you are using a DSLR and want to drift the spectrum, say, 100 pixels (more than enough to "smooth out" the Bayer matrix), then with a 10-μm pixel size this would require an image of 1 mm length.

This would represent an angular distance (assuming a 1,000 mm fl objective) of arctan (1/1,000 mm) = 3.4 min arc. With the RA drive off, the drift rate is 15 min arc/min of time, so this represents an exposure of 3.4/15 = 0.23 min = 14 sec.

In summary, a 14 sec exposure with the drive off would give a spread of 1 mm on the CCD (= 100 pixels) and allow the star to move 3.4 min in arc. The star would still be well within the area of the CCD, so switching the drive back on and hand correcting it back to the starting position would be very straightforward.

In crowded star fields, secondary stars and their spectra can cause problems. If it is not possible to re-orientate the grating to improve the situation, a small aperture stop can be used to suppress this "background noise" (see later).

Camera Response vs. Recorded Spectra

The spectrum obtained will show the effects of the camera CCD response curve. All CCD's have an efficiency curve that normally peaks in the green region of the spectrum. Absorption of the shorter wavelengths (below 400 nm) by the optics together with the low CCD response in this region may limit the extent of the spectrum. DSLR's mated with the Baader type UV-IR filter will give a high efficiency at Hα but will limit the spectral response to between 400 and 700 nm.

These factors cause the spectrum to appear "cut off" in the blue and "fishtail" towards the red. The spectrum will also vary in brightness across its length. During pre-processing, the camera response curve can be calculated and corrections made to spectrum image (see later).

Colored images of the spectrum may look nice, but monochrome images contain as much (if not more) useful data. After checking and rotating the spectrum horizontal (if necessary!), crop around the zero order image and the spectrum, allowing, say, 10–20 pixels above and below the spectrum.

For RBG one-shot color cameras (or DSLR), convert the image to black and white and save the image as a 16-bit file (This format is required if you wish to do further processing). Use pixel binning in the Y axis to "spread" the spectrum, then re-crop a Section 10–20 pixels high. Slight processing with a high pass filter will bring out more detail. Be cautious, though. Over applying filters can add artificial detail and ruin the integrity of the original.

A quick look at the X profile will show the dips in the spectrum caused by the hydrogen Balmer series. See Fig. 7.3.

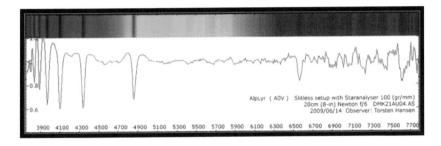

Figure 7.3. Typical spectrum, Vega. (T. Hansen.)

This pre-processed image of Vega can be used to compare the spectrum with similar published data in books and on the web. Quite quickly you can acquire a whole series of spectra that will show the development of the spectral classifications from the blue O-type through to the much redder M-type. Detailed calibration and analysis of spectral images is covered later in this book. See Fig. 7.4.

Using Other Transmission Gratings

Transmission gratings can also be purchased mounted in 35 mm slide frames, or various other sizes. The grating itself is usually protected by thin glass plates on each side and may or may not be blazed. Some of these "educational" gratings may be of low quality and efficiency and should therefore be treated with caution. Cokin produces a 55 mm camera filter, B40-Cosmos, which, although having a very low efficiency, is actually a very usable 240 l/mm grating. Current suppliers are listed in Appendix A of this book.

All of these variations can be used in a similar manner to the 1.25″ filter type detailed above.

Stars with spectral types earlier than the Sun

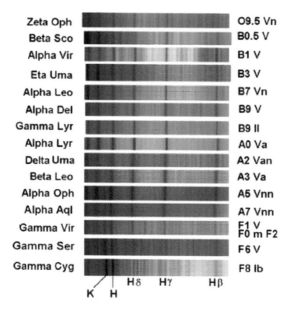

Zeta Oph	O9.5 Vn
Beta Sco	B0.5 V
Alpha Vir	B1 V
Eta Uma	B3 V
Alpha Leo	B7 Vn
Alpha Del	B9 V
Gamma Lyr	B9 II
Alpha Lyr	A0 Va
Delta Uma	A2 Van
Beta Leo	A3 Va
Alpha Oph	A5 Vnn
Alpha Aql	A7 Vnn
Gamma Vir	F1 V / F0 m F2
Gamma Ser	F6 V
Gamma Cyg	F8 Ib

Hδ   Hγ   Hβ

K   H

Figure 7.4. Spectral atlas. (R. Hill.)

NOTE: Where the physical size of the grating surface/holder is larger than the actual area of the grooved grating surface, an aperture stop (a piece of black card with a central opening to match the dimensions of the grooved area) should be fitted to prevent extraneous light from getting through and contaminating the spectrum.

A support frame for mounting these gratings in the converging beam can be made from a 1.25″ T thread nosepiece, a nylon guide block, and a T thread adaptor (Details are given in a later chapter). Depending on the size of the grating it may even be possible to temporarily mount it with masking tape into a 2″ nosepiece adaptor or a filter wheel.

Using Gratings in a Collimated Beam

Transmission gratings (and Amici prisms from a D-V spectroscope) can also be used behind a collimating lens. This can be as simple as positioning it behind a low power eyepiece and focusing the spectrum through a standard camera lens. Unfortunately there are few off the shelf adaptors available to hold the grating onto the eyepiece, and as the grating l/mm increases so does required angular offset of the camera and lens.

There are a couple of DIY options available: The grating can be mounted in a 1.25″ nosepiece T thread adaptor and the camera held it place with an Afocal camera adaptor (These are usually used to image with "point and shoot" cameras through the eyepiece). A low power eyepiece with long eye relief gives best results (One benefit of this method is the opportunity to set the camera at the deviation angle and improve the focus). Drape a dark cloth over the focuser and camera to keep out extraneous light. Alternatively, the grating can be positioned behind the eyepiece in an eyepiece projection adaptor and a T thread filter ring adaptor used to mount the camera lens. This is suitable for 100–200 l/mm gratings where the deviation is only a few degrees. See Figs. 7.5 and 7.6.

Figure 7.5. Eyepiece projection with filter grating, showing the components.

This arrangement has the potential to increase the resolution of the grating. It is still limited by the size of the star image and the optical aberrations. Rigel Systems supply a 600 l/mm grating, mounted in a holder as their "RS Spectroscope," which fits over many of the popular standard eyepieces (Televue, etc.). They also supply a camera adaptor that allows a DSLR with its lens to be mounted securely to image the spectrum.

Incorporating the transmission grating into a slit spectroscope design is discussed later in this book.

Figure 7.6. Final assembly of eyepiece projection with filter grating.

Web Pages

http://astrosurf.com/aras/staranalyser/userguide.htm
www.rigelsys.com

CHAPTER EIGHT

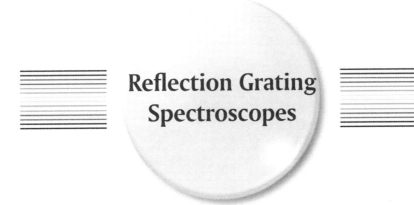

Reflection Grating Spectroscopes

The next significant step in getting higher resolution spectra takes us into the realm of the larger self-contained spectroscopes, usually with an entrance slit and reflection grating.

The option is to use a commercial unit or build your own suitable spectroscope from standard components. See Fig. 8.1. The challenges moving into a larger spectroscope are:

- fitting it securely to the telescope
- matching the collimator f/ratio
- weight and bulk of the instrument
- finding the target and putting it on the slit, and
- guiding the telescope to maintain the image on the slit for the duration of the exposure.

This section covers the more common problems encountered and provides general assistance to all users. All the commercial instruments come with comprehensive instruction manuals, and obviously these need to be read and understood.

There are four commercial astronomical spectroscopes currently (2010) available to the amateur. These are:

1. Baader's Dados-Littrow Spectroscope
2. SBIG SGS Self-Guiding Spectrograph (Ebert-Fastie)
3. SBIG DSS7 Deep Space Spectrograph (Classical)
4. Shelyak LhiresIII-Littrow Hi-Resolution Spectroscope

K.M. Harrison, *Astronomical Spectroscopy for Amateurs*, Patrick Moore's Practical
Astronomy Series, DOI 10.1007/978-1-4419-7239-2_8,
© Springer Science+Business Media, LLC 2011

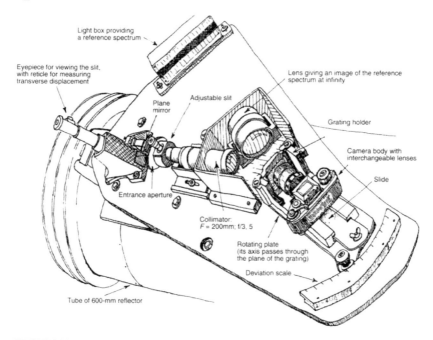

Figure 8.1. Amateur built classic spectroscope, circa 1980. (D. Bardin.)

There's also the Qmax from Questar, but this is only suitable for obtaining solar spectra, and the Optomechanics 10 C, which sometimes becomes available in the secondhand market.

The SBIG units are designed to be used with the SBIG CCD cameras. The Shelyak and Baader spectroscopes can be used with any CCD/DSLR camera, and all use different combinations of focal length collimating/camera lenses to give a range of dispersions/plate scale. Other than the DSS7 all have interchangeable and micrometer-controlled gratings to give low or high resolution spectra. The DSS7 uses a fixed position grating to give a low resolution spectrum. Table 8.1 refers.

Mounting the Spectroscope

The interface between spectroscope and telescope optics is usually the focuser. Many amateurs use SCT telescopes for spectroscopy. The almost unlimited back focus allows secure mounting of equipment using the 2″ SCT threads, and the focal ratio can be readily modified using standard reducers. On other refractors and reflectors 2″-size focusers are preferred, not to prevent vignetting but to provide as rigid a

Table 8.1 Slits and gratings of commercial instruments

Instrument	Slit width (μm)	Gratings (l/mm)	Scale (Å/mm)
Dados	25/35/50	200/900	397/106
DSS7	50/100/150/200	300	600
SGS	18/72	150/600	478/119
Lhires III	15/19/23/35	150–2400	333/13

support as possible. Crayford focusers, with additional locking screws, work well. Also the available travel of the focuser needs to be considered. Like positioning a CCD or DSLR at prime focus, the back focus requirements of the spectroscope nose-piece to the slit position can vary, and this needs to be accommodated. Spectroscopes weigh much more than cameras, and this can cause the focuser tube to sag and the mechanism to slide or move as the telescope tracks. It is much better to work with a minimum extension of the focuser tube and make up any back focus requirements with screwed spacer/extension pieces. See Table 8.2.

Table 8.2 Design parameters of commercial instruments

Instrument	Design f/ratio	Nosepiece fitting	Camera fitting	Weight (kg)
Baader Dados	f10	2″/T thread	T thread	0.85
SBIG DSS7	f10	SCT/T thread	ST 7/8	0.7
SBIG SGS	f6.3/f10	SCT/T thread	ST-7E/8E	0.7
Lhires III	f10	2″/SCT	T thread	1.6

Depending on the telescope, it may also be possible to attach an additional safety strap between the spectroscope and the OTA. A piggyback camera bracket on an SCT makes an ideal anchor point!

Matching the telescope focal ratio to the spectroscope is almost mandatory. Any mis-match will cause loss of throughput and efficiency. Barlow lenses (or better still, one of the TV Powermate, or Baader Telecentric "focal extenders") and reducers can be used to modify the focal ratio. Check the back focus requirements of the extender/reducer; some are designed to function at specific distances (i.e., Meade × 0.63 reducer needs 110 mm back focus).

Consider also how you will see what's going on in the field of the slit. We need to be able to unambiguously locate the target star (or galaxy, etc.) and then ensure that it is positioned on or near the center of the slit. The easiest way is to use a flip mirror in front of the spectroscope. Once the flip mirror has been set up to give a par-focal image on the spectroscope slit and the guide eyepiece, the object can be centered on the cross-wires and verified before using the spectroscope.

All the commercial spectroscopes use fixed-width reflection slits (other than the SBIG DSS7). This allows the "overspill" light from the star to be focused on a guide eyepiece or guide camera for long exposure guiding. The available slit widths vary; see Table 8.1.

Many amateurs have successfully made their own slit mechanisms that tend to be adjustable but not fully reflective. See later for further details.

As part of the initial setup of the spectroscope, the imaging CCD/camera needs to be accurately focused on the slit. Some spectroscopes (Baader Dados, SBIG SGS, etc.) include a small red LED backlight to illuminate the rear of the slit, making focusing easier.

Getting a Star Focused on the Slit

Sounds easy, but when you are looking through a 20 μm slit there's not much to see! It's actually a three-stage process; focus the CCD camera on the slit, focus the guide camera on the slit, and then focus the star image on the slit.

The spectroscope should first be set up to focus the CCD imaging camera onto the slit before mounting it onto the telescope. This can usually be done in ambient light on the bench. The camera should be attached and the grating rotated to bring the zero image of the slit into the middle of the field of view. Adjust the camera mounting to get a clear focused image of the slit, and align it with the edge of the frame. If possible, lock the camera in position to prevent any movement or rotation.

If the spectroscope has a reflection slit, then insert the guide camera and adjust its position until the slit gap is in clear focus. You can also use a flip mirror to position the star. Position the spectroscope on the telescope and orientate the slit along the RA axis. This seems to be the preferred direction; any errors in drive rate just move the star image along the slit, giving a slightly wider spectrum rather than moving off the slit and losing light. With a good guide camera and software it is possible to maintain the star position within fractions of a pixel.

Adjust the position of the grating to view the zero order and set the telescope onto the target star. A well aligned finder and/or a flip mirror is very useful!

Reflection slits will show a "split" image of the star in the guide camera; adjust the telescope focus until you get the smallest star image on the slit. The imaging camera should also show the zero image of the star, and you can use your imaging software to monitor the FWHM size of the image. When you have the smallest possible star image positioned on the slit, everything is ready to start collecting starlight for your spectrum.

Beamsplitters and Non-reflective Slits

Homemade spectroscopes need the same care in focusing on the star image. Using an external guide camera is not an option if there's no star image to guide on!

Off-axis guiders and piggyback guide scopes can be used to give guiding control. The target star must be placed on the slit prior to acquiring a guide star. Remember that the guide star doesn't have to be the target star; any suitable field star close by will do the job. As an alternative, a beamsplitter can be mounted ahead of the spectroscope. These are available in 50:50 split, i.e., 50% of the light goes to the guide camera and 50% goes to the spectroscope. Losing half the available light, almost a magnitude in brightness, is a big price to pay, but the ability to see the whole field of view and be able to select and position the target star on the slit and guide at the same time has some merit. Beamsplitters can be constructed using semi-reflective mirrors or pellicle films, which improve the ratio to 80:20.

Here is your acquisition sequence:

1. Set telescope and focus target object on entrance slit.
2. Rotate grating to center-selected wavelength (if required).
3. Check focus and orientation of spectrum in imaging camera.
4. Take at least two spectrum of neon reference. Switch off and remove from beam.
5. Take several spectra of the target (multiple exposures to build up a long exposure of 10 min or longer).
6. Take at least two spectrum of neon reference. Switch off and remove from beam.

This sequence can be repeated for all the targets during the observing session. At the end of the session take some calibration images:

- dark frame for spectra images
- dark frame for neon spectra
- bias frame

A flat field frame should be taken for every new grating setting or change to the optical arrangement.

SBIG Software Controlled Spectroscopes

SBIG have integrated the design of their spectroscopes with their ST7/8/9 CCD cameras. In the SGS spectroscope (first introduced in the late 1990s) the guide chip "sees" an image at the front of the reflective slit with the target star; the main imaging chip is focused on the rear of the slit and the emergent spectrum. See Fig. 8.2.

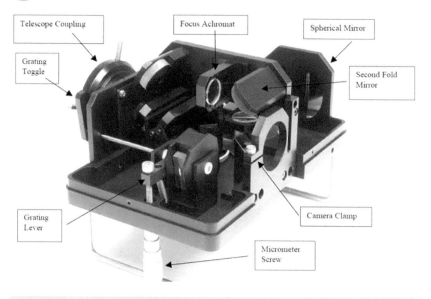

Figure 8.2. SBIG SGS spectroscope. (Courtesy SBIG.)

The camera is controlled through the SBIG CCDOPS camera software package, which has an option when beginning self-guiding to track-to-the-cursor or to-the-centroid of the star. Tracking to the cursor enables the user to reposition the star slightly to straddle the slit. The SGS contains two gratings on a rotating carousel which allows the selection of low and high spectral dispersion at the flip of a lever. The rotation of the grating is controlled via a micrometer head. The Ebert design uses a small spherical mirror for both collimation and camera imaging. Focus is achieved by moving the mirror.

To make the adjustment of the camera orientation easier it's suggested by users that a small hole be drilled in the cover above the camera locking screw. This allows final adjustment to be carried out with the spectroscope on the telescope. SBIG also supply an integrated spectral analysis program, "Spectra," which will process images from the camera cropped to less than 768 pixels wide and exactly 20 pixels tall.

The other SBIG spectrograph is the DSS7, which is unique among the available spectroscopes in as much as it has a fixed low dispersion grating and a motorized slit plate and grating. (The dispersion of the DSS-7 is 600 Å/mm). See Fig. 8.3a.

The optical layout is of the classical design with a doublet collimating and camera lens, and accommodates the SBIG ST-7/8/9/10/2000 cameras (with no filter wheel attached), and the ST-402/1603/3200 cameras. See Fig. 8.3b. The slit plate actually has a selection of slit gaps, a pair of 400 μm slits at the outside, then a 100 μm and 150 μm pair, and finally a single 50 μm slit in the middle. This gives multiple spectra on the imaging camera.

(a)   (b)

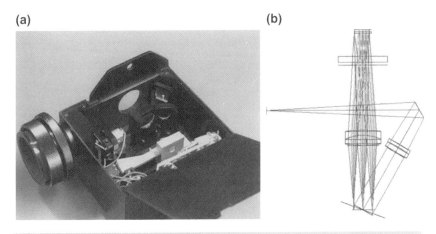

Figure 8.3. SBIG DSS7 spectroscope. (Courtesy SBIG.)

To focus the instrument the slit plate is rotated into the optical path and the grating rotated to zero order position. Both of these movements are controlled through the CCDOPS software. With the slit plate visible the internal camera focusing lens can be adjusted to give best focus and alignment. This lens also acts like an ×0.5 reducer and gives a wider field of view and effectively reduces the projected width of the smallest slit to 25 μm.

At the telescope, the slit plate is rotated out of the optical path and the grating rotated to zero position to provide a full field view of the target area. The target star can be positioned on a "virtual" slit using the software and the grating/slit plate rotated back to its normal operating position. This allows the camera to record a full frame image of the spectrum.

Baader Dados Spectrograph

Baader's spectroscope design looks a little odd, being that it is two cubes joined together! This configuration (Dados= Spanish for dice) was developed by Baader in 2008, in collaboration with the Max-Planck Institute and CAOS, a group of professional spectroscopists at the European Space Agency (ESO).

The first cube contains the telescope nosepiece, entrance slits, viewing port for guiding, and an illuminator. The second holds the collimating lens, rotating grating, adjusting micrometer, camera lens, and the imaging camera connection. See Figs. 8.4a and 8.4b.

The reflective slit plate provided has three slits: 25, 35, and 50 μm width and is front illuminated.

The imaging camera attachment provides a collection of adaptors that allow visual eyepieces, webcams, DSLR's, and CCD's to be quickly fitted and aligned.

(a)   (b)

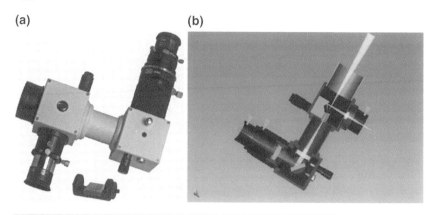

Figure 8.4. The Baader Dados spectrograph. (Courtesy Baader.)

Focusing on the zero order slit is aided by the front illumination, which also helps to frame the target star while positioning it on the slit gap.

The guide port is 1.25″ and has a transfer lens built in. This can be adjusted to give about 150% magnification if required. The 200 and 900 l/mm gratings (25 mm × 25 mm) are easily changed over to give a wide range of dispersion/plate scale. A nice feature is that the grating can be locked in position after adjustment.

Shelyak's LhiresIII

This spectroscope was originally designed and developed by a group of French amateur astronomers in the 1990s and was initially offered as a DIY kit. It was very popular and became the foundation of many spectroscopic observing programs in Europe. Shelyak Instruments refined the design and brought it to the commercial market in 2005. See Figs. 8.5a and 8.5b.

The LhiresIII has an f10 Littrow configuration (but can be used between f8 and f12) with a 200 mm focal length doublet. A 12 V power supply (not required for normal use) is used for the neon reference lamp.

The reflective entrance slit is selectable from 15/19/23/35 μm to suit the size of the star image. The neon reference lamp is built in to the nosepiece, and it can be swung in front of the slit during calibration. A 1.25″ guide port and transfer lens provide guiding capability on the slit.

The grating support platform is available with a selection of gratings (The 2400/1200 l/mm are 25 mm × 50 mm and the 600/300/150 l/mm are 25 mm × 25 mm). This gives a very wide range of dispersions and plate scales.

After attaching the imaging camera, the grating is focused on the slit by means of a small helical focusing ring (via an access side panel) which moves the collimating lens forward and backward.

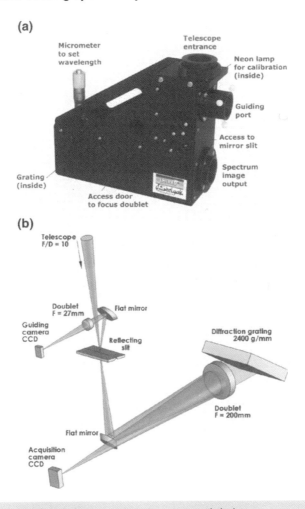

Figure 8.5. The Shelyak LhiresIII spectroscope. (Courtesy Shelyak.)

Questar QMax Solar Spectrometer

This instrument was developed by Questar Corp. in 2000 for use on their range of premium 3.5″ Maksutov telescopes. It is a dedicated solar spectroscope and cannot be used to obtain stellar or DSO spectra. Primarily a visual instrument using a Brandon 8 mm eyepiece it can show a 68 Å wide section of the solar spectrum with a

Figure 8.6. The Questar QMax spectrometer. (Courtesy Questar Corp.)

dispersion/plate scale of 11.6 Å/mm and a resolution of 0.17 Å. The grating rotation has a small digital readout in 2 Å per division. See Fig. 8.6.

Built-in UV and IR filters protect the user from excessive radiation. The spectrum can be imaged through the eyepiece using the Afocal method.

Other Spectroscopes

Also available to amateurs are the new generation of echelle spectroscopes. Both Shelyak and Baader have models that became available in 2009.

The eShel Spectrograph from Shelyak

The new "eShel" echelle spectroscope consists of three modules and support software. The Fiber Injection and Guide module (FIG) is designed to be mounted on an f6 telescope (typically an SCT f10 with a × 0.63 reducer) with T threads and has a perforated (50 μm aperture) mirror to reflect the overspill starlight to the guide camera. There's also an electromagnetic flip mirror that, when energized, allows the light fed by a 200 μm fiber optic (from either the calibration lamp or a white LED) to enter the entrance aperture. These lights and control signal for the flip mirror come from the calibration unit. A 3 m 50 μm optical fiber connects the FIG to the body of the spectroscope. See Fig. 8.7.

The calibration unit contains the high voltage power supply for the thorium-argon reference lamp and a flat field white LED light illuminator. This LED also

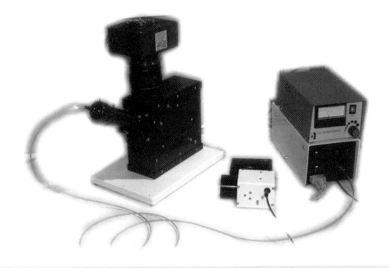

Figure 8.7. The Shelyak eShel spectrograph. (Courtesy Shelyak.)

assists in identifying the position of the various spectral orders in the field of the imaging camera during the calibration process.

The spectrograph module has an inlet port for the 50 μm fiber optic cable and a 125 mm f5 collimating lens. The echelle grating (79 l/mm, blazed at 63°) is positioned at an incident angle of 63.45° and reflects the composite order spectrum at 5.75°. An AR coated prism then acts as a cross disperser to present some 20 spectral orders (#32–#52) to the imaging lens. Shelyak has used a Canon 85 f1.8 lens as the imaging lens, allowing both DSLR's and large frame CCD's to be used (It is optimized for a KAF1603 chip 13 mm × 9 mm and 9 μm pixel). With this lens the various spectral orders are spread across the CCD chip and are approximately 12 pixels wide with a 6 pixel background gap between them.

The software developed by Shelyak, *eShel*, has been designed to fully automate the collection and analysis of the spectrum. It can use the white light flat field image to identify the position of the spectral orders on the CCD frame and assist in the identification of the reference light emissions. The resolution, $R = >10000$ over a spectral range of 430–700 nm, is impressive.

The Baader Baches Echelle Spectroscope

The Baader "Baches" weighs only 2 kg and mounts onto directly onto the focuser via a 2″ nosepiece or T thread adaptor. Designed for an f10 entrance beam it is fitted with a fixed 25 μm ×100 μm slit. See Fig. 8.8.

The 79 l/mm, 63° blaze angle echelle grating combined with a 300 l/mm cross dispersion grating gives an image of 29 spectral orders (#32–#61) on a APS-sized

Figure 8.8. The Baches spectroscope. (Courtesy Baader.)

CCD chip, and coverage from 350 to 780 nm. The resolution, $R = 18000$, gives a theoretical resolution of < 0.3 Å

A thorium-argon lamp is used for calibration, and MIDAS Echelle software is used to analyze the spectrum. An exposure of 15 min will record a spectrum of a 5 mag star with an SNR of 50.

The Optomechanics 10 C

Many of these spectrographs, produced by Optomechanics Research, Inc., are still in use and occasionally come up for sale on forums such as AstroMart. The basic design from the 1970s is very robust, and it became the standard spectroscope for many colleges and universities. See Fig. 8.9.

Model 10 C has interchangeable reflection 50 and 100 μm slits ruled on aluminized glass, 1.5 mm long with comparison openings at both ends, and built-in reference lamps (Hg-Ne, Fe-Ne, or Fe-A hollow-cathode). The light is introduced via optical fibers permanently mounted in the spectrograph. A spherical mirror (225 mm f9) acts as the collimator, and the 10 C has interchangeable plane gratings (from 300 to 1,200 l/mm) and a Nikon 135 mm f2.8 camera lens for imaging. The weight of the unit is approximately 4.5 kg.

Remote Control of Spectroscopes

More and more amateurs are using their instrumentation remotely, or semi-remotely. The use of planetarium software, such as the freeware *Carte du Ciel,*

MODEL 10C ASTRONOMICAL SPECTROGRAPH

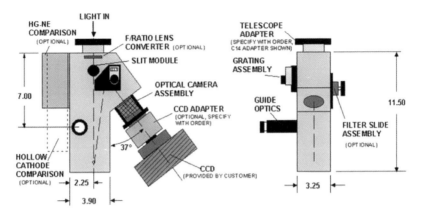

Figure 8.9. The optomechanics 10 C spectrograph. (Courtesy R. Hilliard.)

and the advent of quality GOTO mountings such as the Skywatcher NEQ6 pro series combined with the strong support from groups such as ASCOM/EQMOD allow the amateur to set up local networks (or even just USB/serial hubs) to provide very accurate telescope GOTO and positioning capability. Focus of the imaging camera and sequencing the exposures can be achieved nowadays with precision USB/stepper motor focusers, and most of the commercial imaging processing software packages will allow the sequencing (darks/flats and images) of camera exposures. Spectroscopy has the added complication of star acquisition onto the slit, reference lamp exposures, and adjustment of the grating position. The challenges can be overcome by using suitable flip-mirror/guide camera combinations and USB devices to switch on and move the reference lamp onto the entrance slit and rotate the grating (if required).

Web Pages

http://www.televue.com/engine/page.asp?ID=42
http://eq-mod.sourceforge.net/
http://ascom-standards.org/
http://www.stargazing.net/astropc/
http://www.skywatchertelescope.net/swtinc/product.php?class1=3&class2=304
http://www.socastrosci.org/2008%20papers/2008SASc..27..103C_Paper.pdf

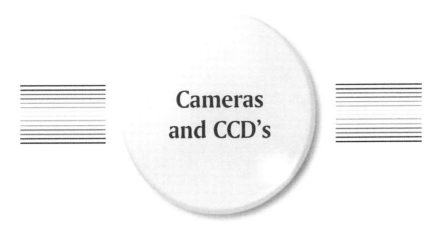

CHAPTER NINE

Cameras and CCD's

Any of the modern CCD cameras and DSLR's can be successfully used for spectroscopy. The use of modded DSLR's if fitted with UV-IR cut off filters (i.e., Baader) will be restricted to wavelengths between 400 and 700 nm.

The main characteristics of the CCD chip that impact its effective use are:

- CCD chip size
- pixel size
- quantum efficiency

CCD Chip Size

Sizes vary, from 3.6 mm × 2.7 mm (typical webcam) to the latest APS-C, which is 22.7 mm × 15 mm. When used with different dispersion/plate scales obviously the larger CCD will allow more of the spectrum to be recorded during the exposure.

Pixel Size

Based on the Nyquist sampling criteria of a minimum 2 or 3 pixel image, the spectroscope's final resolution can be limited by the size of the pixels in the imaging camera. To record faint spectra using 2×2 binning may help to improve the sensitivity, but at the cost also of reduced resolution. Try and match the pixel size to the slit/collimating/camera lens capabilities. See Table 9.1.

K.M. Harrison, *Astronomical Spectroscopy for Amateurs*, Patrick Moore's Practical
Astronomy Series, DOI 10.1007/978-1-4419-7239-2_9,
© Springer Science+Business Media, LLC 2011

Table 9.1 CCD chips

CCD chip	QE (% age)	Pixel size (μm)	Pixel array	Chip size (mm × mm)	Resolution Nyquist (μm)
ICX098BQ	??	5.6	640×480	3.6×2.7	11.2
ICX249AK	70	8.6×8.3	752×580	6.5×4.8	17.4
KAF-0401E	50	9.0	768×512	6.9×4.6	18
ICX085AL	50	6.7	1,300×1,030	8.7×6.9	13.4
KAF-0261E	??	20.0	512×512	10.2×10.2	40
KAF-1602E	50	9	1,536×1,024	14.0×9.3	18
KAF-3200ME	60	6.8	2,148×1,472	14.8×10.2	13.6
EOS 300D	30	7.1	3,072×2,048	22.7×15	14.2
Nikon D40	60	7.8	3,032×2,016	23.6×15.7	15.6

Due to the Bayer matrix in DSLR chips, a minimum of 4 pixels need to be illuminated to record all the incoming wavelengths of light. The final resolution will be slightly less than a mono CCD chip and will depend on the debayering model used.

Quantum Efficiency

Each sensor has a unique quantum efficiency curve, sometimes known as a sensitivity curve. Some chips will give a usable transfer efficiency (say, above 20%) from the UV through to the IR, with a peak in the green of around 550 nm. Most chips show a maximum quantum efficiency below 60% (NOTE: Some literature shows the curves as "Relative response." These can be misleading in as much as they do not give an actual QE figure but rather a percentage of the maximum). Figures 9.1, 9.2, 9.3, 9.4 and 9.5 show typical response curves.

In the case of one-shot color (OSC) chips (or DSLR's) the RGB response curves can show significant dips in efficiency between the various color filters used. When using a DSLR always save images in RAW format. This maximizes the bit depth and avoids compression artifacts.

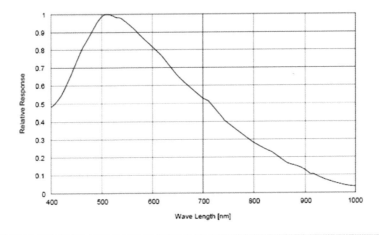

Figure 9.1. QE curve for ICX098BL-DMK. (Sony.)

Bayer Matrix

In all color cameras the color response of the individual pixels are controlled by the use of a colored filter array (CFA) called the Bayer filter (after Bryce E. Bayer of Eastman Kodak who developed and patented it in 1976). This consists of a series of red, green, and blue filters (RGB) or cyan, magenta, and yellow (CMY) placed in a matrix over the CCD pixels, usually 50% green, 25% red, and 25% blue. All the

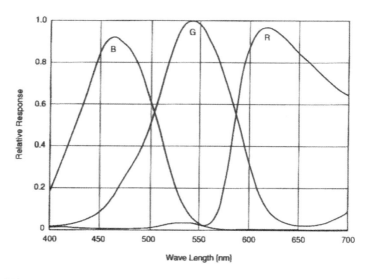

Figure 9.2. QE curve for ICX098BQ-color webcams. (Sony.)

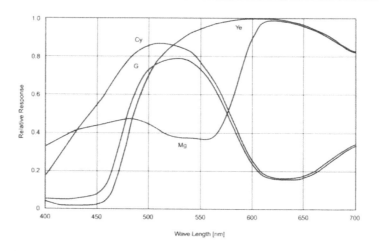

Figure 9.3. QE curve for ICX429AKL DSI II. (Sony.)

DSLR's use the RGB as well as the color SBIG cameras, Starlight H8C, H9C, M25, etc. The DSI, DSI II, Starlight MX7C, and so on use the CMY. Effectively each pixel records a monochromatic intensity signal of the incident light. See Fig. 9.6.

To achieve a color image output from the various pixels the Bayer matrix must be "translated" back into an RGB color. Various debayering algorithms have been developed. Some, like the Bilinear method, give low resolution compared with the

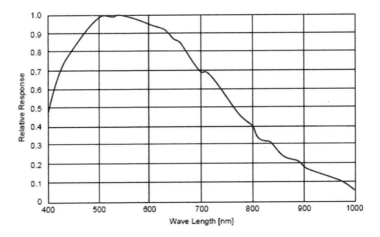

Figure 9.4. QE curve for ICX285AL DSI III/SXV-H9. (Sony.)

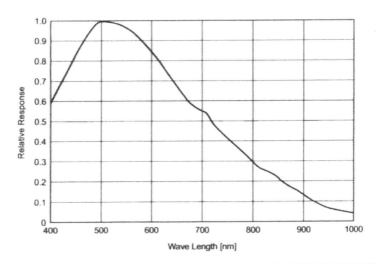

Figure 9.5. QE curve for ICX424AL-atik16ic. (Sony.)

more sophisticated VNG (variable number of gradients) model. Dr. Craig Stark has written a very good article on the subject (See Web Pages at the end of this chapter). The positioning of the spectral image on the Bayer matrix can significantly affect the resolution and response of the CCD.

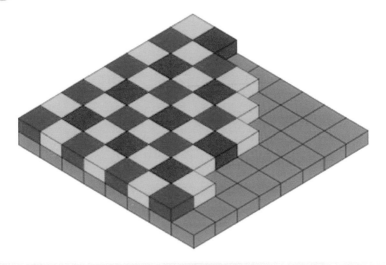

Figure 9.6. Bayer matrix. (WIKI.)

It's recommended therefore that the spectrum always be positioned as close to the horizontal axis of the chip as possible. Images saved as RAW files give the maximum dynamic range.

Bias, Darks, and Flats

Like conventional astrophotographs, the image of the spectrum will also contain spurious data (noise) from the CCD chip. As the final spectrum is only a few pixels wide any noise or hot pixels can distort the subsequent analysis.

Noise signals in the CCD image are made up of a "base" noise associated with the electronic transfer of the image from the CCD chip, which doesn't change with exposure duration or temperature. Further "system" noise is generated by temperature (thermal noise) and hot pixels. We also have distortions introduced by the optical system; vignetting, internal reflections, and dust.

A short-duration dark exposure "bias" frame records the "base" noise, a "dark" exposure, i.e., an exposure of the same duration as the normal exposure but with the optics covered (for example, no external light getting in) will compensate for the "system" noise, and a "flat" exposure, a short-duration exposure of a uniformly illuminated field, will show the vignetting, and so on.

By subtracting these bias, dark, and flat images from the actual spectral image we can correct for the majority of noise. These processing steps are covered in any of the popular books on astrophotography.

Spectroscope Flats

Normally, as we've already seen, any light entering the spectroscope will record as a spectrum image whether we are using a filter grating or a slit spectroscope; the issue is to how convert this to a usable flat image.

There are two aspects to the solution. The first is to use a light source that generates a continuum spectrum, i.e., no emission or absorption lines, and the second is to take multiple flats across the whole spectrum by rotating the grating as necessary. A suitable light to use is a quartz halogen lamp (150 W), white high intensity LED's, or an electroluminescent panel. These have a very uniform spectrum. Directed to a sheet of white paper/white T shirt draped over the telescope objective this light will give a usable flat. See Fig. 9.7.

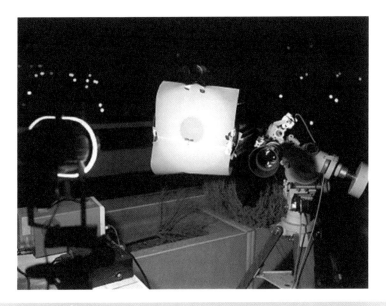

Figure 9.7. Taking a flat with a diffuser and halogen lamp. (C. Buil.)

A small diffuser (white paper/white acrylic) positioned inside a black cardboard tube sized and positioned for the focal ratio of the telescope (i.e., if the telescope is, say, an f6 focal ratio – a 20 mm circular mask placed 120 mm from the slit will give you the same entry beam) can be used instead of mounting the spectroscope on the telescope. Much more convenient! See Fig. 9.8.

Depending on the dispersion of the spectroscope, several flats centered on different wavelengths may be required to give good coverage. See Figs. 9.9, 9.10 and 9.11. Remember, a flat taken in the red region will not be usable as a correction flat image in the blue region. Any change to the spectroscope configuration (i.e., change

Figure 9.8. Taking a flat with illuminated diaphragm. (C. Buil.)

Figure 9.9. A flat frame image. (C. Buil.)

Figure 9.10. Image before flat subtraction. (C. Buil.)

Figure 9.11. After flat subtraction. (C. Buil.)

of telescope focal ratio, grating, camera distance, camera lens, or CCD) will require new correction images.

Further Reading

Covington, M. A. *Astrophotography for the Amateur.* Cambridge University Press (1985).
Covington, M. A. *Digital SLR Astrophotography.* Cambridge University Press (2007).
Ireland, R. S. *Photoshop Astronomy.* Willmann Bell (2005).
Wodaski, R. *The New CCD Astronomy.* New Astronomy Press (2002).

Web Pages

http://www.licha.de/astro_article_ccd_sortable_compare.php
http://www.astrosurf.com/buil/iris/tutorial5/doc17_us.htm
http://astrosurf.com/aras/lhires_flat/flat.htm
http://www.stark-labs.com/craig/articles/assets/Debayering_API.pdf
http://www.cloudynights.com/item.php?item_id=2042
http://www.cloudynights.com/item.php?item_id=1973
http://www.cloudynights.com/item.php?item_id=1966
http://www.cloudynights.com/item.php?item_id=2001
http://www.eaas.co.uk/news/astrophotography_resources.html
http://tech.dir.groups.yahoo.com/group/ccd-newastro/files/CCDCalc%20Install%20Files/
http://home.arcor.de/l.schanne/Einsteiger/Flats/flats_e.htm

CHAPTER TEN

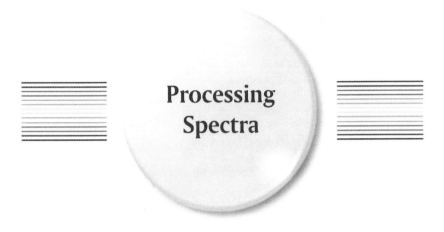

Processing Spectra

Once you have obtained the raw spectral image the next stage is to analyze the spectrum and extract as much useful data as possible. You'll be amazed with the information hidden in that small strip of light! With care and the right software it is possible to determine stellar temperatures, chemical constituents, and sometimes the physical nature of the star.

Note: Most of the analysis software available to the amateur calculates wavelengths in Angstroms. This is easily converted, if needed, to nanometers (nm) by dividing the Å value by 10, i.e., 1000 Å = 100 nm.

Some basic analysis can be done using profiles exported from your imaging program and then imported into a spreadsheet (such as Excel). SBIG, as mentioned earlier, has its own integrated spectra software that allows calibration to be done. The most popular freeware programs for pre-processing and analyzing spectra are IRIS, SPIRIS (a subset of IRIS developed for the Lhires III spectrograph), and Visual Spec. There is also Sky Spec, an analysis program that handles spectra obtained using gratings (i.e., linear spectra). Others, such as IRAF and MIDAS, which were originally used by professionals running under Linux and are now available for Windows; these are highly sophisticated programs and as such have a very steep learning curve!

There are a series of steps that need to be taken in analyzing the spectrum. These include:

- preparing the raw image for processing.
- obtaining a pixel profile.
- wavelength calibration.
- identification of other lines and features.

K.M. Harrison, *Astronomical Spectroscopy for Amateurs*, Patrick Moore's Practical
Astronomy Series, DOI 10.1007/978-1-4419-7239-2_10,
© Springer Science+Business Media, LLC 2011

Preparing the Raw Image for Processing

The spectral image should be corrected using bias, dark, or flats before subsequent processing. Use your imaging program to correct and align the spectrum horizontal across the image, within one or two pixels (A vertical height of only eight or ten pixels is required for analysis).

When using multiple exposures (some faint target stars may require a total exposure of up to 30 min) use a stacking program such as Deep Sky Stacker (DSS) or Registax to combine the images. See Fig. 10.1.

Figure 10.1. Vega spectrum taken with a SA100 and modified Canon DSLR.

Some sharpening of the spectrum can be done with hi-pass filters, but be cautious; too much sharpening can add artifacts that will destroy the integrity of the image. Your raw image, if in color, needs to be converted to black and white. This can be done easily enough using your imaging software by extracting the luminance channel (L) or combining the RGB data into a 16-bit Fits file.

The freeware image processing software called IRIS V5.58 and developed by Christian Buil can also do this (and a lot more); the drop down tab "Digital photo"/48–16 bits converts a color image to a 16-bit black and white. This can then be saved as a Fits file (or a .PIC file for VSpec – see later).

Grating Spectra: Linear Dispersion

Spectra obtained with a grating are basically linear and can easily be converted from pixels to wavelength if at least two reference points are known. For filter gratings, the zero order image can be taken as one reference; this only leaves a pronounced line feature (one of the Balmer lines for instance) to be identified (or guessed!) to calibrate the spectrum in Angstrom. Spectra obtained from prisms unfortunately need a multiple reference point calibration due to their non-linearity.

A cursory investigation of the spectrum can be achieved by using a profile taken through the spectrum. See Fig. 10.2.

Importing the data file from this profile into an Excel spreadsheet allows a basic wavelength calibration to be carried out. A couple of definite lines need to be identified. Based on the difference in wavelength between the lines and the number of pixels between them on the image you can calculate an Å/pixel calibration factor.

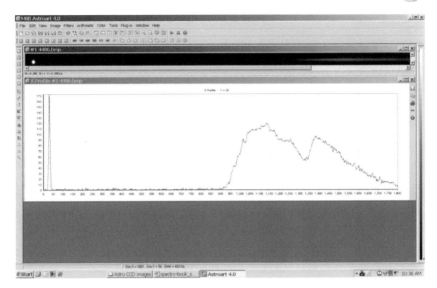

Figure 10.2. Spectral profile Vega ex-AA4.

For instance if your first line is Hβ at 4861 Å and the pixel number is 830, and the second identified line is, say, Hα at 6563 Å and the pixel number is 2,175, the calibration would be (6,563–4,861)/(2,175–830)=(1,702)/(1,345) = 1.26 Å/pixel.

Multiply the pixel number by the calibration factor (i.e., pixel 830 would now represent 1046 Å, etc.). The scale is now correct; we just need to register an off-set for wavelength. The pixel at 1046 Å actually equates to a wavelength of 4861 Å, so the off-set would be +3,815 (4,861-1,046). When this off-set value is added to each pixel value the result will represent calibrated wavelengths in Å. If a zero order image is available in the frame, the calibration is simplified by using only one reference line and the zero order.

This data can then be used to generate a wavelength calibrated graph, which will show the spectral profile. See Fig. 10.3.

Smile and Tilt

The optical arrangement of some slit spectroscopes can give rise to distortions of the spectral lines. In some configurations the spectral lines can appear curved and tilted relative to the axis of the spectrum. IRIS has built-in commands to correct these distortions.

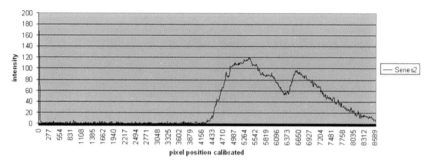

Figure 10.3. Excel calibrated profile graph.

Visual Spec (VSpec)

Due to its ease of use and strong support base, the freeware "Visual Spec" by Valerie Desnoux has become the "de-facto" standard software for amateurs' processing spectra. It has a built-in library (based on Pickles data) of spectrum profiles for all the stellar classifications, a very complete library of elements and their emission lines, integrates with R. O. Gray's "Spectrum" software (a freeware synthetic spectrum generator), and can export spectral data to allow further processing in spreadsheets.

The input file, an image of the spectrum, must be in either a black and white 16-bit Fits format or the proprietary.PIC format. Pre-processing by applying darks and flats and aligning the spectrum horizontally across the frame should be done prior to importing into VSpec. The spectrum must also have the blue region of the image towards the LHS. The image can be flipped if necessary.

Standard Stellar Spectra

Vspec contains a listing of spectral type for all the stars in the *Bright Star Catalogue*, down to approximately 6.5 magnitude. Under "Tools/Spectral Type," type in the Bayer letter or the Flamsteed number of the star and constellation, i.e., Betelgeuse, α Orionis would be entered as "AlpOri"; this brings up "Spectral Type" – M1-2Ia-Iab. This can then be used to find a suitable comparison spectrum under the VSpec Library. Open "Tools/Library" and in the drop down screen look for a matching spectrum i.e. M2Ia.dat.

Note that VSpec uses spurious prefixes (1 and 2) to order the listing of spectra. It also uses the following prefixes:

wf – metal weak
rf – metal rich

Standard Element Lines

There is also an extensive listing of elemental lines available for comparison within VSpec. These can be accessed through "Tools/Elements":

Element.txt, – based on CRC Handbook
Lineindent.txt – ILLSS catalogue of lines in Stellar Objects
Sun.txt – Astronomical Astrophysics. Suppl. Ser. 131, 431
Hires – A database in 38 sections based on R. O. Gray's "Spectrum"
Atmos – Listing prepared by C. Buil from various sources

Obtaining a Spectral Pixel Profile

By importing a black and white 16-bit Fits file (or a VSpec.pic file) of our spectrum, we can quickly produce a 1 D profile that will show the intensity variations for the spectrum by pixel location.

If the image has been properly pre-processed we can use the "object binning" command. This adds all vertical intensity values in each pixel column across the image; otherwise an area of the spectrum can be selected using the "reference binning zone" icon to provide a new binning area.

Click the "object binning" icon, and a new screen will be generated showing the profile of your spectrum (note this also changes the file name from.FIT to a.SPC). The top header panel shows the position of the red cursor line (L) and the intensity value (I) as the cursor is moved across the profile. Note that at this stage the title "Å/pixel" reads 0.

Look carefully at the shape of the profile curve, and see if there are any noticeable dips that could represent hydrogen absorption lines. These are the easiest to use for calibration. Don't worry too much at this stage what the overall shape of the curve looks like; concentrate on the features.

In the example, Fig. 10.4, we have a spectral image of Vega taken with a Baader filter grating and a modified Canon 300 D camera (internal filter replaced with a Baader UV-IR filter).

The out of focus zero image is on the left hand side. The spectrum shows some prominent dips (at the cursor position), and we will assume that it is the Hβ line of hydrogen at 4861 Å. With this uncalibrated profile we can move on to the wavelength calibration process.

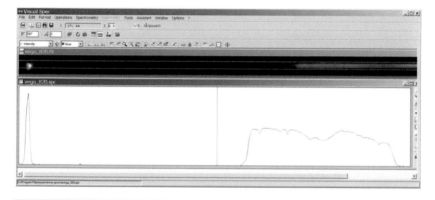

Figure 10.4. Vega raw image.

Wavelength Calibration

The easiest method to calibrate an unknown spectrum is to use two reference points (One can be a zero image if the filter gratings are used). Click on the "Calibration two lines" icon; a message screen will appear the first time this is used, which asks for confirmation to make the profile an Intensity Series; confirm by clicking Yes. This will generate a new profile file that can be used for subsequent operations. The header portion of the screen now changes to Fig. 10.5.

Position the cursor just to the left of the zero image, press and hold the LH mouse button, and drag the cursor to the RH side of the zero image; this will generate two dotted lines and a box containing a number; highlight the number and type in

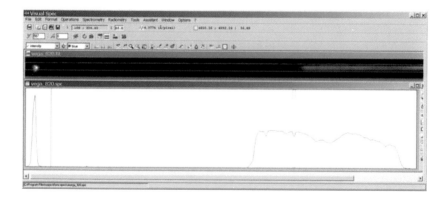

Figure 10.5. Vega calibration screen.

"0000." This will appear in red and is the wavelength of the zero image. Press Enter to confirm. The box will disappear but the dotted lines will remain.

Look up at the header panel and you will now see against the title "ie 1," a box with "0000." This is the first reference line.

Repeat this for the dip in the profile and type in "4861." The Å/pixel title now shows 4.3774. This is the calibration/plate scale of the profile. Note also as you move the cursor over the profile that the numbers in the L window change – these reflect the wavelength and intensity of the point on the profile below the cursor.

Now we have a calibrated profile, we can note the plate scale 4.3774 Å/pixel, this can be used in future to assist in subsequent calibrations using the same grating/camera set-up. We can also use some of the other VSpec tools. Under the tab "Tools/Library" a window will open with a listing of stellar spectra from O types through to M types (Note the first 1o5 V.dat is an artificial number to get the listing in order! It's actually an O5 V spectrum).

Click to highlight "a0V.dat" and hold the button down, then drag the highlighted file onto the profile screen and a second profile (in purple) will appear. This is the library spectrum profile of a A0V star similar to Vega. Straightaway you will see the registration between the dips of the original spectrum and those of Vega. See Fig. 10.6.

If we now return to the "tools" and "elements" in the second window click "H" for hydrogen and press "sort." This will list the wavelengths of the hydrogen emission lines in the profile; press "export" and a series of orange lines will appear on our profile. See Fig. 10.7.

Note how these match the dips in the Vega reference spectrum and the dips in the original. We can now confirm the hydrogen absorption lines in the original as Hγ, Hβ, and Hα at 4340.47, 4861.33, and 6562.72 Å, respectively. Note also how the

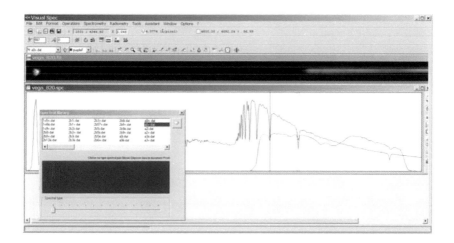

Figure 10.6. Vega profile with A0V reference spectrum.

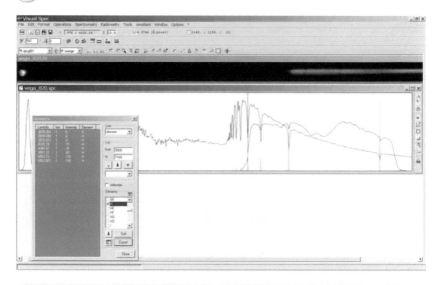

Figure 10.7. Vega profile with comparison profile and H lines.

profile drops off sharply at 4000 and 7000 Å. This is due to the Baader UV-IR filter fitted to the camera.

Camera Response

In the example Vega profile, it is obvious that the shape of the recorded spectrum profile does not match that of the reference star. The difference is due to the varying response of the camera over the wavelengths recorded. It is possible to use the reference star profile to generate a camera response that can be used to correct future spectra taken with the same setup. If you can access the CCD chip QE curves you can also prepare a response curve as a.DAT file for importing into VSpec.

Select a range of wavelengths to be used for the camera response. This will usually be around the 3900–7000 Å, which is close to the QE limits for most cameras. Although there may appear to be data outside these limits, the very low camera response can cause spurious results. It also suppresses atmospheric line around 7200 Å. Unless you require data from these regions for a specific task it's best to cut them out of the profile.

Using the "crop" icon, click the LH button at around 3900 Å; this will show a dotted line. Hold the button and move the cursor across to 7000 Å; release and the second dotted line will show the extent of the selected range.

Save this profile. Open "Tools/Library/" and highlight a0v.dat; drag the file onto the new cropped profile. It will appear as an orange profile. See Fig. 10.8.

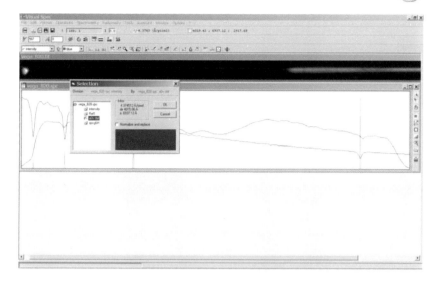

Figure 10.8. Cropped raw profile with reference profile.

We now need to divide the original profile by the reference profile. Select the active profile in the top screen left hand box as "Intensity," Open "Operations/Divide profile by profile," and select aOv.dat from the drop down box.

This will generate a new green profile, "Division." See Fig. 10.9.

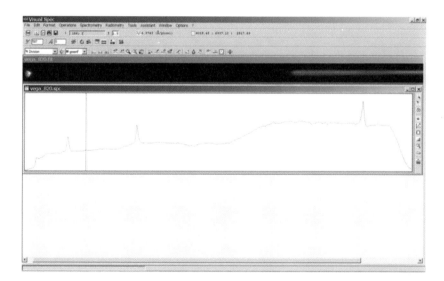

Figure 10.9. VSpec preliminary camera response.

You will see that the green profile now shows large "emission" lines where the original profile had the hydrogen absorption lines. To make things easier to process click on the "Erase graphic" icon and select "Division" as the active profile from the list. This will clear the profile window and re-load the green profile for further processing. VSpec only works with "Intensity" profiles, so the division profile needs to be converted to an Intensity profile. This is done through "Edit/Replace" and in the drop down box. Select "Intensity" and press Enter. Our green profile will now turn blue, and the selection box will show Intensity. Save this file as xxxxx_division.spc, or similar.

To convert this profile into a usable camera response curve we need to smooth the curve and remove the "emission lines." This is easily done with the "Radiometry/Compute Continuum." Clicking it will bring up another line of icons and change the profile from blue to orange. The first icon "point/courbe" allows the selection of points along the profile that will be used to generate a continuum curve. Using this command we can suppress the "emission lines" (and any other line features) to get a consistent curve. See Fig. 10.10.

Click on as many points as you need on the profile, obviously missing the lines, then press the "execute" icon. This will open a drop down screen "Continuum – coefficient" and show a new orange profile corresponding to the points selected.

The continuum co-efficient should be selected to give a smooth continuous curve, and can be altered by editing the initial value "20" to around 2,000; the slider then will show a value of 0–2,000 as it is moved down the scale. Very high values (experiment with a value >5,000) will make the curve too smooth and will not accurately reflect the camera response curve. When you're satisfied with the result click OK on the drop down screen and the profile window will change to show the final

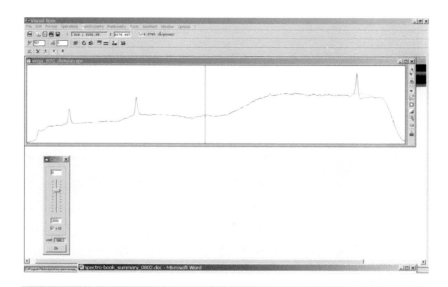

Figure 10.10. VSpec smoothed camera response curve.

result. Delete the contents of the display window and just bring up the "fit.intensity" profile. Use "Edit/Replace" to make it an active profile and save as a response.spc (Choose a name to suit your camera and setup).

Correcting Spectra Using the Camera Response Curve

This response curve can be used to correct spectra taken under similar conditions, i.e., same telescope/spectroscope/camera. With a new profile open, select a similar spectral region to that recorded on the camera response, open the response profile, and use "Edit/Copy"; click on the new profile window to use "Edit/Paste" to place it on the new profile. You should now see the new profile in blue and the response curve in purple. Click on the profile box and select "Intensity" as the active profile. Using "Operations/divide profile by profile" select the response Intensity curve. The new green profile will show the result. Again the star spectrum and element lines can be copied from the library to show the calibration. See Fig. 10.11.

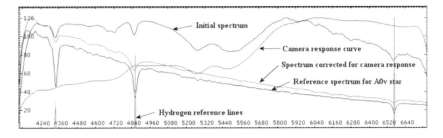

Figure 10.11. Profiles showing raw profile, camera response, reference spectrum, and corrected profile.

Using the CCD QE Curves – Response Curve

By measuring the QE curve for your CCD, a.dat file can be prepared and used in VSpec. The required format needs a continuous list of two columns:

Column 1 – Wavelength in Å
Column 2 – Intensity (or, in this case, QE values)

The Intensity column must contain at least one non-zero value. This file can be prepared either in Excel or as text file – save as text and rename as .dat. Once renamed, this file can then be imported into VSpec and used normally.

Calibrating Using a Reference Lamp

Spectra taken with a slit spectroscope can also be calibrated using a standard reference lamp.

There are a couple of ways of doing it; with the same camera and without moving the grating take an exposure through the slit of the reference lamp or using a Dekker type screen and expose the lamp at the end of the spectral exposure. The first method relies on registering the pixel positions between the images to obtain an accurate calibration, whereas the second superimposes an image of the reference lamp on the same image.

The Lhires III has a built-in neon reference lamp to allow taking a reference exposure. The SBIG SGS has an external window to allow a reference lamp to shine onto the front of the slit. Any distortions or tilt in the image must be removed before processing.

From the reference lamp exposure a reference profile can be generated through VSpec (NOTE: VSpec assumes that the reference image is exactly the same width in pixels as the image being calibrated).

First, open the target image and bin to show a profile curve. Now open the reference lamp image, and this time use the "reference binning" to prepare a profile that will show as "Ref1." Open this reference profile and calibrate. For a neon light, the line listing detailed in the VSpec elements file may be useful in determining the wavelengths of the various lines.

Resolution

One method of establishing the resolution of the spectroscope is to measure the full width half maximum (FWHM) of an emission/absorption line in the spectrum. VSpec can do this calculation automatically based on a smooth continuum on either side of the line (or better still, after the spectrum has been normalized). Select the "Spectrometry/computation preferences/" tab and click the FWHM box.

Normalized Spectrum

This function, "Operations/Normalization," is used when comparing equivalent widths of lines taken at different times, or when stacking profiles for display and visual analysis. The resulting profile shows the relative intensities with the continuum as the base, equal to one.

Signal to Noise Ratio (SNR)

VSpec can provide an approximate value for the SNR in a spectrum. This is based on the ratio of standard deviation to the average. For best results this should be done on a number of line free sections of the continuum. A SNR of above 50 is good; for a ProAm contribution a SNR of >100 would be acceptable.

Continuum Removal

To make line features easier to see it is sometimes preferable to remove the continuum from the profile and effectively "flatten" the curve.

First normalize the profile by using a similar process to that for camera response; points are selected on the continuum line, a profile is generated and smoothed, and it is then divided into the target profile.

Equivalent Width (LEQ)

The equivalent line width is an important parameter in using line profiles for analysis of quantities of atoms (element abundance) in stellar atmospheres (curve of growth), Doppler spread, etc. LEQ can be calculated from a selected line in a normalized profile using "Spectrometry/computation preferences/" and ticking the LEQ box.

This is just a brief introduction to the capabilities of VSpec. Read though the various tutorials that cover processing in much more detail than the above examples.

Web Pages

http://www.sc.eso.org/santiago/uvespop/bright_stars_uptonow.html
http://deepskystacker.free.fr/english/index.html
http://www.astronomie.be/registax/
http://astrosurf.com/aras/spiris/spiris.htm
http://astrosurf.com/aras/spiris_en/spiris_en1
http://www.astrosurf.com/vdesnoux/
http://www.astrosurf.com/buil/us/iris/iris.htm
http://astrosurf.com/buil/iris/tutorial10/doc27_us.htm
http://www.eso.org/sci/data-processing/software/esomidas//midas-overview.html
http://iraf.noao.edu/

CHAPTER ELEVEN

Amateur Spectroscope Projects

There are many satisfying projects that can be undertaken with the spectroscope; some are more suited to the higher resolution slit designs, but the majority can be attempted with even the basic transmission filter grating. Internet forums and groups that are open to all amateurs interested in astronomical spectroscopy are very useful; there you'll find that the more experienced forum members are only too willing to share their knowledge and assist the novice.

Solar Spectrum

Any slit spectroscope or a transmission grating using a reflective needle can be used to image the Sun's spectrum. Use extreme caution when using the spectroscope on a telescope to view the Sun; the objective should be well stopped down. A 4 mm aperture will provide more than enough light to get good results (Remember also to cover the finder!). Imaging the solar spectrum will not add to our knowledge of the Sun but will allow to you get practice in using your spectroscope, i.e., setting it up to the telescope, focusing, grating position calibration, determining camera response, and estimating plate scale and resolution. Analyzing the solar spectrum will also show the effects of the camera response curve and the changing focus position between the blue and red regions. See Fig. 11.1.

Indentifying lines in the solar spectrum is made easier with the collection of solar spectral data available on the web (See the Web Pages listed at the end of this

K.M. Harrison, *Astronomical Spectroscopy for Amateurs*, Patrick Moore's Practical
Astronomy Series, DOI 10.1007/978-1-4419-7239-2_11,
© Springer Science+Business Media, LLC 2011

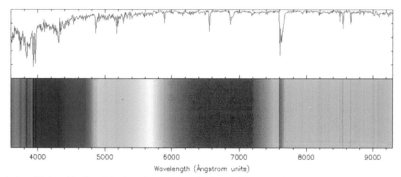

Data from "Photometric Atlas of the Solar Spectrum from 3000 to 10,000 A" by L. Delbouille, L. Neven, and C. Roland
Institut d'Astrophysique de l'Universite de Liege, Observatoire Royal de Belgique, Liege, Belgique, 1973
Image copyright © 2002 by Ray Sterner, Johns Hopkins University Applied Physics Laboratory

Figure 11.1. Liege atlas – full solar spectrum.

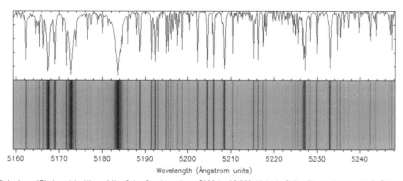

Data from "Photometric Atlas of the Solar Spectrum from 3000 to 10,000 A" by L. Delbouille, L. Neven, and C. Roland
Institut d'Astrophysique de l'Universite de Liege, Observatoire Royal de Belgique, Liege, Belgique, 1973
Image copyright © 2002 by Ray Sterner, Johns Hopkins University Applied Physics Laboratory

Figure 11.2. Solar spectrum – *Mg lines.*

chapter). The Liege atlas is recommended, as it allows various wavelengths to be seen in detail and provides a fully annotated identification of lines. See Figs. 11.2 and 11.3.

Astronomical Filters

The bandwidth and transmission factors for all the available astronomical filters can be analyzed using a slit spectroscope. The filter to be tested should be placed in front of the slit and illuminated by a "flat" spectrum source, i.e., a quartz halogen lamp (or even the Sun, for this type of experiment). The bandwidth of the filter will be

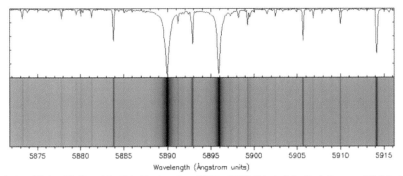

Wavelength (Ångstrom units)

Data from "Photometric Atlas of the Solar Spectrum from 3000 to 10,000 A" by L. Delbouille, L. Neven, and C. Roland Institut d'Astrophysique de l'Universite de Liege, Observatoire Royal de Belgique, Liege, Belgique, 1973 Image copyright © 2002 by Ray Sterner, Johns Hopkins University Applied Physics Laboratory

Figure 11.3. Solar spectrum – *Na lines.*

obvious, and the shape of the spectrum will give an indication of the transmission. See Fig. 11.4. For more precise measurements a calibration spectrum of a neon reference can be used. After calibration, various elements and their lines (i.e., hydrogen, oxygen, mercury, sodium, etc.) can be brought in from the VSpec library for comparison.

Another easy way of checking the effectiveness of the light pollution rejection type filters is to set up the telescope and spectroscope pointing towards a light pollution glow (or street lights) and take a couple of spectral images; put the filter in front of

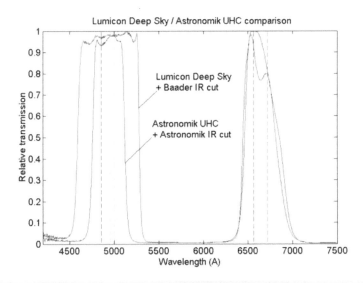

Figure 11.4. Comparison spectral profiles for Lumicon and Astronomik filters. *Dotted lines* from the *left* are: Hβ, OIII (doublet), Hα, and SII. (C. Buil.)

the slit and retake an image. The differences are obviously due to the bandwidth and efficiency of the filter!

Stellar Classification

With a transmission grating filter (i.e., Rainbow Optics/Star Analyser/Baader) or an objective prism/grating, it is possible to replicate the work done by the team at Harvard and prepare your own detailed stellar classification atlas. See Figs. 11.5 and 11.6. A selection of bright stars covering the B- to M-types will provide comparison spectra with an exposure of a minute or so. A good starting point would be:

Acrux, αCrucis –	B0.5IV
βCentauri –	B1III
Adhara, εCanis Majoris –	B2II
Spica, αVirginis –	B2V
Achenar, αEridani –	B3V
Regulus, αLeonis –	B7V
Vega, α Lyrae –	A0V
Sirius, αCanis Majoris –	A1V
Castor, αGeminorum –	A1V
Deneb, αCygni –	A2 Ia
Altair, αAquilae –	A7V
Canopus, αCarinae	F0II
Procyon, αCanis Minoris –	F5 IV
Sun –	G2V
Rigel Kentaurus, αCentauri –	G2V
Capella, αAurigae –	G5III
Pollux, βGeminorum –	K0III
Arcturus, αBootis –	K1III
Aldebaran, αTauri –	K5III
Antares, αScorpii –	M1.5Ib
Betelgeuse, αOrionis	M2Ia
Mira, oCeti -	M7IIIe

Torsten Hansen has also obtained many stellar spectra using the SA100 grating on a 200 mm f6 Newtonian telescope. See Fig .11.7.

Jack Martin, using a Rainbow Optics 200 l/mm grating, black and white film, and a 250 mm Newtonian has completed and published a comprehensive spectral atlas of the brighter stars. See Fig. 11.8.

Emission Stars (Be and WR)

Be stars, or "shell stars," are complex stars where high spin rates eject matter that forms a ring around the star. This ring emits emission lines of hydrogen that are

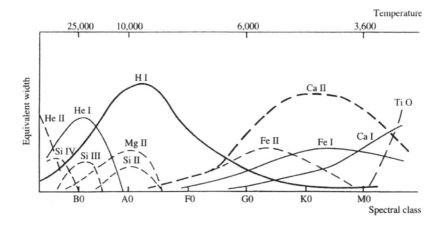

Figure 11.5. Equivalent *line* widths. (Struve, O. *Elementary Astronomy*, p. 259, Cambridge, 1959.)

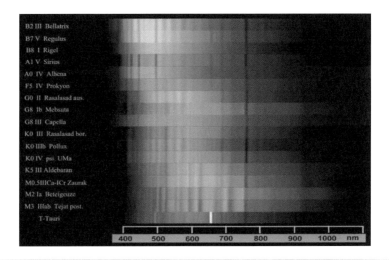

Figure 11.6. Stellar classification using SA100. (R. Bahr.)

broadened by the Doppler shift and are seen superimposed on absorption lines from the star's atmosphere. They therefore appear doubled, with a bluer approaching wing and a reddened receding wing. For amateurs this phenomenon is more noticeable at the Hα line as the "P Cygni" profile – i.e., AX Persei, AB Aur, BX Monocerotis. See Fig. 11.9.

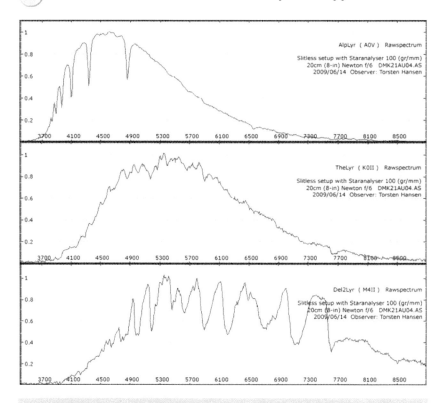

Figure 11.7. Spectral profiles of Alpha, Theta, and Delta Lyrae. (T. Hansen.)

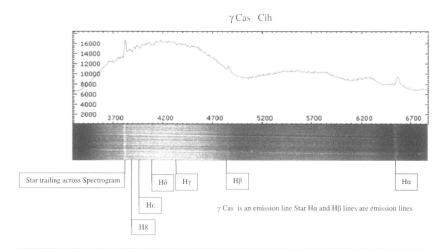

Figure 11.8. Gamma cass spectrum. (J. Martin.)

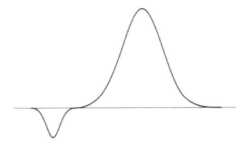

Figure 11.9. P Cygni intensity profile. (K. Robinson.)

Wolf Rayet (WR) stars (first discovered in 1867 by Charles J. E. Wolf (1827–1918) and Georges A. P. Rayet (1837–1906)), are extremely hot O-type stars that show numerous emission bands of hydrogen, ionized helium, carbon, nitrogen, and oxygen, all associated with the enormous mass transfer going on at their surface. Matter is streaming away from the star so that we only see the spectrum of the lower density out-flowing material. The WR classification is further divided into WC (carbon emission stars) and WN (nitrogen emission stars). Some 15% of all O- and B-type stars show emission spectra. See Fig. 11.10.

A catalog of over 2,000 Be stars and some 46,000 spectra from over 500 stars is available at the Be Star Spectra (BeSS) website.

Doppler Shift – Binary Stars/Exoplanets/Quasars

One of the first applications of the spectroscope was to resolve close double stars by measuring the Doppler shift of their spectra (The first to be discovered, during preparation of the Draper catalog, was ζ UMa, Mizar in 1887. It showed a 2 Å shift in the spectral lines).

As the stars rotate around their common center of gravity they will appear to be approaching or receding from us; this leads to an apparent doubling of the spectral lines (Spectral lines can also appear doubled by Earth's annual rotation, where it may be approaching or receding from the star at up to 30 km/s). Christian Buil on his website has a nice animation showing the Doppler effect in the Hα line of the 4.7 magnitude binary double star 57 Cyg. His calculations give a period of 2.8548 days, and a maximum radial velocity of 450 km/s.

Fulvio Mete has recently measured the Doppler shift of the binary star Beta Aurigae, recording a velocity of 100 km/s. See Fig. 11.11.

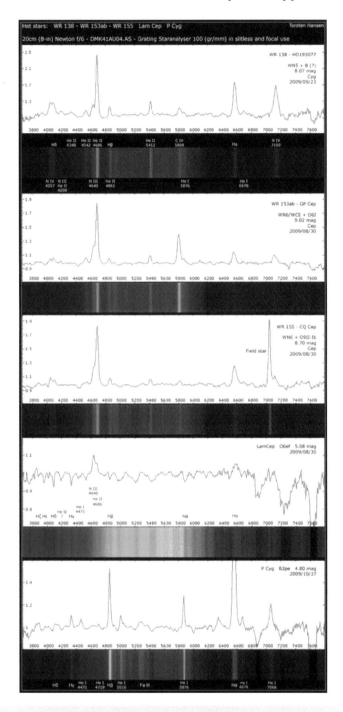

Figure 11.10. Selection of WR spectrum. (T. Hansen.)

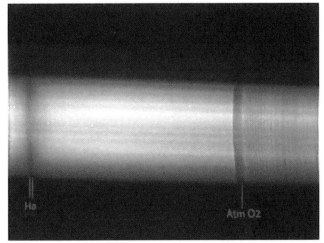

Beta Aurigae Spectrum - October, 26 , 2009 h.22 UT -
Ponte di Nona Observatory -Rome
Celestron 14 @ 11- COGOS (COncave Grating Opticsless Spectroscope)
Atik 16 HR CCD camera , binning 2 x 2.   - Fulvio Mete

Figure 11.11. Double spectral *lines* in Beta Aurigae spectrum. (F. Mete.)

Obviously, high resolution spectroscopes can record smaller Doppler shifts. Tom
Kaye heads up a worldwide network of observers under the banner of "SpectraShift
– Extrasolar Planet search project."

Maurice Gavin was the first amateur to record the redshift of a quasar in 1998.
He successfully determined the redshift (z) for a 13th magnitude quasar, 3C273, as
0.16. This work was repeated by Christian Buil in 1999 when he obtained a result of
$z = 0.156 +/- 0.015$. See Fig. 11.12.

Variable Stars, Nova, and Supernovae

Spectral observations of variable stars and novae can complement the traditional
amateur work being done visually and photometrically in plotting the light curves
of variable stars.

Changes in the relative intensity of the emission lines can be seen to parallel the
light changes of the stars as observations of cataclysmic variables such as SS Cygni
show. Other stars of interest are: HD 34959, V378 And, o Ceti (Mira), R Leonis, and
R Cygni.

Recently (2009) Robin Leadbeater obtained some significant spectra of the binary
star ε Aurigae that demonstrated that the long-awaited eclipse was about to begin.
These eclipses occur only every 27.1 years. See Fig. 11.13.

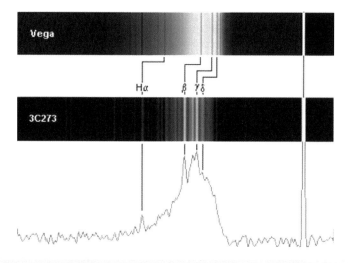

Figure 11.12. Spectrum of 3C273 showing redshift compared with Vega. (M. Gavin.)

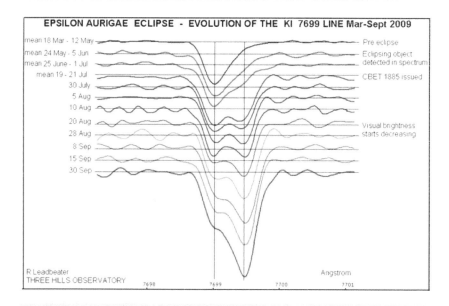

Figure 11.13. KI absorption *line* variations in e Aurigae. (R. Leadbeater.)

Novae (and some recurring novae) occur when a white dwarf star erupts. The brightness of the star can increase by up to 10 magnitudes over a matter of a day and then gradually decline over a few months. The spectrum can be very complex, at maximum brightness looking like an A-type supergiant. During decline absorption

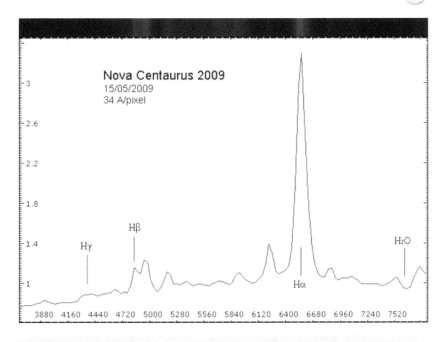

Figure 11.14. Spectrum of nova centaurus 2009 – SA100 grating. (T. Bohlsen.)

lines are seen. Then a transition to broad emission lines more typical of gaseous nebulae are seen. See Fig. 11.14.

Supernovae (SN) can appear in any one of the thousands of galaxies visible in amateur telescopes. There are two basic types: Type I, which has a rapid rise in brightness followed by a low, slow decline; and Type II, which has a much broader plateau following maximum brightness. Type Ia shows no hydrogen lines during brightening whereas Type II does. The fast-evolving broad-lined (caused by the high expansion rates) Type Ic SN has an extremely short rise time. The typical discovery magnitude of 15 mag puts them into the difficult category, but Christian Buil and others have successfully obtained spectra with filter transmission gratings. See Fig. 11.15.

The Harvard Smithsonian Center for Astrophysics (CfA) maintain a large database of SN spectra (over 700 of 49 stars) going back to 1972 and cover all SN types (Ia, Ibc, and II).

Nebulae

There are three basic types of nebulae observed by amateurs: bright emission nebulae, such as M42 in Orion; bright reflection nebulae, such as the glowing surround of M45 Pleiades; and planetary nebulae, such as M27, the Dumbbell.

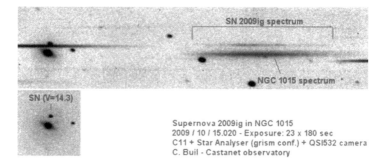

Figure 11.15. Spectrum of SN 2009ig. (C. Buil.)

The emission nebula emit discrete wavelengths due to the ionized atoms of hydrogen (Hα, Hβ), helium (HeI 5876 Å), oxygen (OII 3727 Å) OIII (4959 Å, 5007 Å), NII (6548 Å, 6583 Å) and sodium (SII 6717 Å, 6731 Å), whereas the reflection nebula usually show the continuous spectrum of the exciting star nearby.

The planetary nebula are gaseous remnants of stellar collapse and usually emit OIII (4959 Å, 5007 Å), NII (6548 Å, 6583 Å), HeI (5876 Å), HeIII (4686 Å), Hα, and Hβ light.

Huggins was the first astronomer to view emission nebulae (the planetary nebula, NGC 6443, in 1864) through the spectroscope, and his results clearly showed "The answer, which had come to us in the light itself, read: Not an aggregation of stars but a luminous gas." See Fig. 11.16.

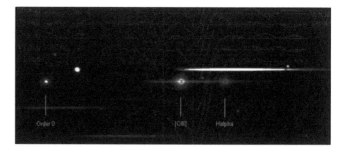

Figure 11.16. NGC 2392 SA100+Grism. (C. Buil.)

When a slit-less spectroscope is used, the emission spectrum shows repeat images of the nebula in the wavelength being emitted. Regions of ionized oxygen can show different images to those of the hydrogen, etc., due to the complex nature of the nebula.

This spectral image (Fig. 11.17) of the Tarantula Nebula in the LMC, taken with a 55 mm camera lens and a SA100 grating, clearly shows the differing shape of the emission regions.

Figure 11.17. Tarantula nebula spectrum. (R. Kaufmann.)

Comets

The typical structure of a comet is the head or nucleus surrounded by the coma, an envelope of gas and dust, and radiating tails either of dust or gas (ion tail). The majority of the comet's light is reflected light from the Sun, with some emission spectra from the excited gaseous compounds inside it. Cometary spectra is dominated by the Swan bands (named after the Scottish scientist William Swan, who mapped them in 1856) of carbon and its compounds (Fig. 11.18), i.e., CH, and CN, mixed with some NH_2 and oxygen.

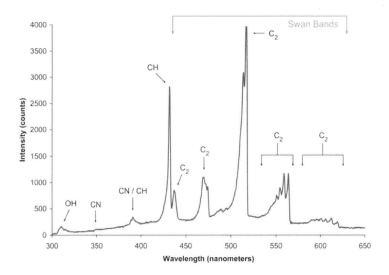

Figure 11.18. Swan bands. (WIKI.)

A slit spectroscope is needed to obtain meaningful spectra of the various parts of the comet. The slit can be positioned on the nucleus and/or the tail to obtain spectra of the gases and solar reflections from the dust particles.

Hα Observing of Solar Prominences

Another first for Huggins was the recording of a Hα solar prominence without the aid of an eclipse. In 1869 Huggins applied his spectroscope, fitted with a red filter and a movable diaphragm close to the slit (to reduce the amount of excess light entering the spectroscope) to the edge of the Sun's disk. By focusing on the region around the Hα absorption line, he noted that prominences appeared as bright areas in the otherwise dark band. By opening the slit he was able to see the whole extent of the prominence.

Amateurs can do the same experiment today using a slit spectroscope. A refracting telescope stopped down to around 50 mm and fitted with a red filter (a Wratten 25A or equivalent) will allow you to examine the edge of the Sun; unfortunately the spectroscope has to be rotated to maintain the slit tangential to the edge of the disk. The slit can be carefully opened and the edge scanned for prominences. If you're lucky and one is found, further opening of the slit will show the full extent.

Special spectroscopes, called spectrohelioscopes, invented in 1924 by G. E. Hale, have been developed to allow a scanned image of the Sun to be obtained in any nominated wavelength. Fred Veio has designed and built many spectrohelioscopes, which have provided excellent results for the last 40 years.

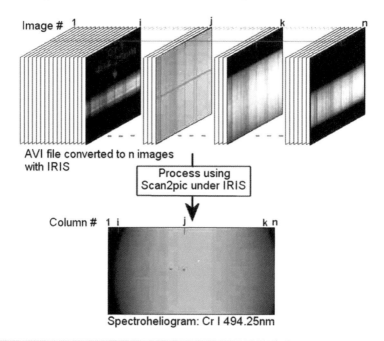

Figure 11.19. Method of scanning the solar image. (After D. Defourneau.)

Recently (2002) Daniel Defourneau developed an electronic spectrohelioscope, where a series of CCD exposures could be combined to present a whole disk image of the Sun in the selected wavelength using his program SpecHelio.exe. See Figs. 11.19 and 11.20.

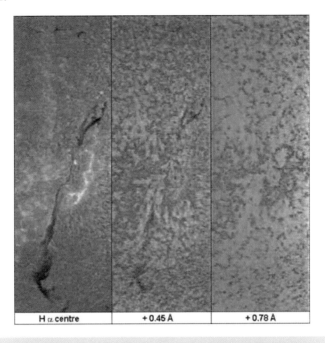

| H α centre | + 0.45 Å | + 0.78 Å |

Figure 11.20. Webcam spectrohelioscope image of an Hα flare. (D. Defourneau.)

Meteors

As a meteor enters Earth's atmosphere the energy released ionizes the particles in both the atmosphere and the meteor to give a plasma emission spectrum. The first photograph of a meteor spectrum was recorded at Harvard in 1897 using an objective prism.

The following emission lines can usually be recorded:

H and K lines of CaII (3934 Å, 3969 Å)
Calcium (4227 Å, 6162 Å)
Iron (4046 Å-4144 Å, 4268 Å-4427 Å, 4886 Å-4988 Å, 5371 Å-5456 Å)
Magnesium (3855 Å)
MgI (5167 Å, 5173 Å, 5184 Å)
Oxygen (5577 Å, 6157 Å)
Sodium (5890 Å, 5896 Å)
SiII (6347 Å, 6371 Å)

Figure 11.21. Meteor spectrum. (E. Majden.)

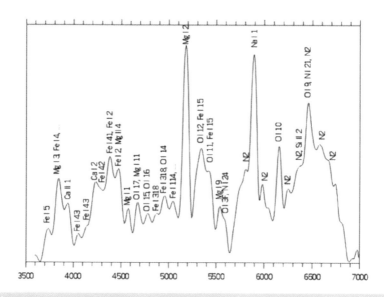

Figure 11.22. Meteor spectrum profile. (E. Majden.)

Similarly, today, an objective prism/grating mounted on a photographic lens will give good results. The grating should be orientated parallel to the anticipated track of the meteor to give best dispersion and allowance made for the deviation angle of the prism/grating.

Ed Majden (BC Canada) has been recording meteor spectra for over 50 years using a 600 l/mm grating and various film cameras (Bronica and Hasselblad) as well as a 50 mm Canon lens coupled to an image intensifier. (Figs. 11.21 and 11.22).

Planetary Spectroscopy

Planets and their satellites shine by reflected sunlight. It is possible, however, to distinguish some of the gaseous compounds in the atmospheres of the larger planets such as Jupiter, Saturn, Uranus, and Neptune (Fig. 11.23). The Swan bands of methane are very prominent and easily recorded with even the SA100 filter grating (Figs. 11.24 and 11.25).

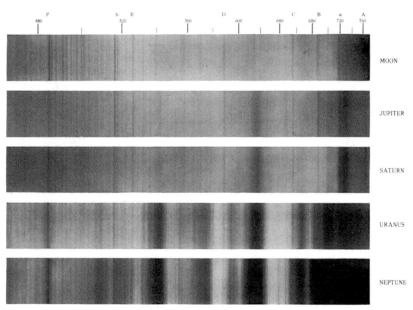

THE SPECTRA OF THE MAJOR PLANETS.

Figure 11.23. Prism spectra of the gas planets compared with the Moon. (Courtesy Lowell Observatory.)

Astronomical Spectroscopy for Amateurs

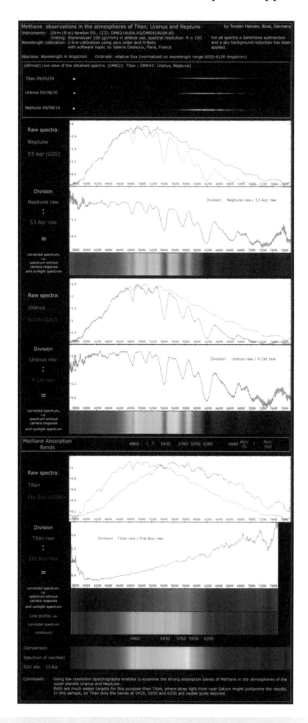

Figure 11.24. Methane spectra – Titan, Uranus, and Neptune. (T. Hansen.)

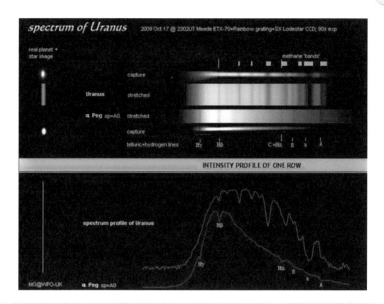

Figure 11.25. Methane spectra of Uranus. (M. Gavin.)

Further Reading

Abt, H. A., Meinel, A. B., Morgan, W. W., Tapscott, J. W. *An Atlas of Low-Dispersion Grating Stellar Spectra.* Kitt Peak National Observatory (1968).

Airey, D. *High resolution spectra and monochromatic images of a flaring 1991 Perseid meteor* J. Br. Astron. Assoc, 109, 4 (1999).

Cox, J., Monkhouse, R. *Philip's Colour Star Atlas E2000.* Geo Philip Ltd (1991).

Karttunen, H., et al. (Ed.). *Fundamental Astronomy, 4th Edition.* pp. 214–216. Springer (2003).

Martin, J. *Spectroscopic Atlas of Bright Stars.* Springer (2009).

Maunder, E. W., Sir W. *Huggins and Spectroscopic Astronomy.* TC & EC Jack (1913).

Philip, A. G. D. (Ed). *Objective-prism and Other Surveys.* L Davis Press (1991).

Webb Society. *Deep Sky Observer's Handbook,* Vol 2. Enslow (1979).

Web Pages

http://fermi.jhuapl.edu/liege/s00_0000.html
http://astrosurf.com/jpmasviel/unpeudescience/saf_ohp_spectro_comete.pdf
http://basebe.obspm.fr/basebe/Accueil.php?flag_lang=en
http://www.peripatus.gen.nz/Astronomy/Index.html
http://www.hposoft.com/EAur09/EAurSpectRef.html
http://adsabs.harvard.edu/abs/1958JRASC..52..169H

http://www.amsmeteors.org/spectro.html
http://members.shaw.ca/epmajden/index.htm
http://www.eyes-on-the-skies.org/shs/
http://www.astrosurf.com/cieldelabrie/sphelio.htm (In French)
http://www.spectrashift.com/
http://www.tomkaye.com/astronomy_extrasolar_planets.shtml
http://www.webbdeepsky.com/wperiodical/article117.pdf
http://www.astroman.fsnet.co.uk/quasars.htm
http://www.astrosurf.com/~buil/us/spe6/quasar.htm
http://www.hposoft.com/EAur09/EAurSpectRef.html
http://www.skyandtelescope.com/observing/home/51804622.html
http://www.threehillsobservatory.co.uk/astro/spectra_40.htm
http://www.astronomie-amateur.fr/feuilles/CV/SS%20Cyg.html#obs
http://astrosurf.com/buil/sn2009/spectra.htm
http://www.cfa.harvard.edu/supernova//SNarchive.html
http://www.lesia.obspm.fr/perso/jean-marie-malherbe/papers/setup.pdf
http://www.unitronitalia.it/baader-dados/dados.htm
http://www.shelyak.com/
http://astrosurf.com/thizy/lhires3/index-en.html
http://www.samirkharusi.net/spectrograph.html
http://astrosurf.com/buil/filters/curves.htm
http://www.astronomie-amateur.fr/Projets%20Spectro1.html (In French)

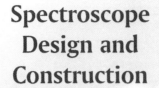

PART THREE

Spectroscope Design and Construction

CHAPTER TWELVE

Design Basics

In Part I the basic theory of how a spectrum is produced by prisms and gratings as well as the various types of prism, transmission, and reflection grating spectroscopes were briefly discussed. Here we will concentrate on the actual design elements of the spectroscope components. This information will give the interested amateur a better understanding of the critical aspects of spectroscope design and how these can be applied to the construction of his/her own spectroscope.

Other than objective prism/gratings and filter gratings in a converging beam application (discussed later), the astronomical spectroscopes designed and built by amateurs generally follow the classical design layout and have the following five main elements:

- entrance slit
- collimator
- dispersion prism/grating
- imaging lens
- eyepiece/film holder (or CCD chip)

To this is added the attachments to the telescope, the guiding system, and the reference lamp.

Figure 12.1 details the general arrangement of a classic spectroscope fitted with an entrance slit. The telescope with aperture D_T and focal length F_t illuminates the entrance slit, width s, which is at the focus of the collimating lens D_1, focal length F_1. The grating, width W, is set at an entry angle α (relative to the grating normal)

K.M. Harrison, *Astronomical Spectroscopy for Amateurs*, Patrick Moore's Practical
Astronomy Series, DOI 10.1007/978-1-4419-7239-2_12,
© Springer Science+Business Media, LLC 2011

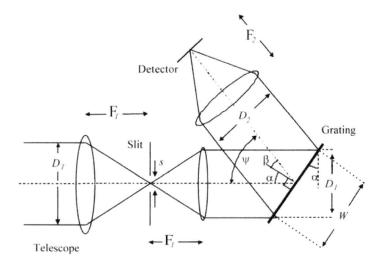

Figure 12.1. Optical arrangement of a classical spectroscope. (After J. Allington-Smith.)

and reflects a dispersed beam D_2 at angle β, which is imaged onto the CCD with a lens of focal length F_2.

The overall optical efficiency of the spectroscope (sometimes called throughput, light gathering power, or etendue) can be as low as 10–15%, due to factors including:

- entrance slit size
- reflections and light loss through the collimating or camera lens
- vignetting of either the collimating or camera lens
- vignetting of the grating
- reflection efficiency of the grating
- effective "blazing"
- order of spectrum being observed
- quantum efficiency of the CCD sensor

We will look at the design of each of the elements before considering how they are manufactured and assembled together to maximize the spectroscope efficiency.

Excel spreadsheets are available to assist in the design of spectroscopes. Here are two that are particularly useful:

- TransSpec.xls provides all the necessary calculations to evaluate a transmission grating in a converging beam.
- SimSpec.xls is a spreadsheet originally prepared by Christian Buil and translated into English (with annotated comments) to assist in the design of a collimated grating spectroscope.

These, together with other data and notes, are available on a Springer website (See Appendix D at the end of this book).

The Entrance Slit

If you read any of the old books on spectroscopy you'd probably find they spend a lot of time talking about slits and how difficult they are to make and adjust. Why are they so important? A couple of reasons: ideally the light must be presented to the grating as a parallel pencil aligned with the rulings on the grating and secondly the spectral image produced by the camera lens is a "picture" of the slit FOR EACH WAVELENGTH. For example, if there were no slit, just a small round opening, then the spectrum seen would be a collage of small colored disks overlapping along the image, making it very difficult to see the fine detail. Generally the finer the slit gap, the better the detail. It is this ability to give clear spectral "lines" that separate the "slit-less" designs, i.e., filter gratings, from the professional spectroscopes.

A slit doesn't have to be linear. During solar eclipses, when the Sun is about to be totally covered by the Moon, a flash spectrum of the corona can be recorded without a slit on the spectroscope. The decreasing arc of the visible Sun acts as a slit, and the emission lines are seen as arcs on the image. However, to assist in identifying and measuring absorption and emission lines the straight slit has become the recognized standard.

The dimensions of the slit gap can vary from instrument to instrument. The width of the slit can be from 20 to more than 50 μm wide (20/1,000 to >50/1,000 of a mm), and the height from 3 to 6 mm (long slit design). For comparison, the average human hair is 70 μm diameter. The main criteria are that the edges of the slit jaws be square, flat, smooth, and parallel. Achieving all these requirements can be a challenge. Also, to assist guiding on a star image all of the commercial spectroscopes use reflective slits. These are made of highly polished stainless steel or etched on a chromed glass plate.

When positioned at the focus of a telescope the slit will be illuminated by the star image. The size of this image is dependent on the focal length (F_t) of the system and the seeing conditions (The star image produced by the telescope has a Gaussian light distribution, where the peak intensity drops off rapidly. The full width half max (FWHM) of this curve is usually taken as the size of the star) (Fig. 12.2).

The linear size of the star image = seeing disk size × plate scale

$$\text{Plate scale} = F_t{}^*\pi{}^*10^3/180{}^*3600 (\mu m/arc\,sec)$$

$$\text{Plate scale} = 4.848{}^*F_t{}^*10^{-3} (\mu m/arc\,sec)$$

Example: for a 250 mm f6 telescope ($F_t = 250{}^*6 = 1{,}500$ mm)

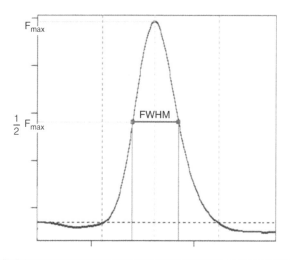

Figure 12.2. Gaussian curve – FWHM.

$$\text{Plate scale} = 4.848^{*}1,500^{*}10^{-3}\,(\mu m/arc\,sec)$$

$$= 7.3\,\mu m/arc\,sec$$

For typical seeing conditions of a 3″ arc (FWHM), this gives a linear star size of 22 μm. If the slit width is greater than 22 μm then the whole star image will pass through, and the spectroscope effectively becomes a "slit-less" design. There have been some investigations as to the effect of narrow slits (and pinhole apertures) on the transmission efficiency of the slit.

The image of a star may be considered as a 2 D Gaussian function, Fig. 12.3, 1a, defined by the FWHM and plate scale.

An entrance slit (narrow slit or pinhole) centered on the star image as seen from the telescope side would appear as shown in Fig. 12.3, 1b (slit) and 1c (pinhole). Figure 12.3, 1d represents the flux coming through the slit and entering the instrument (viewed from the spectroscope collimator), while Fig. 12.3, 1e represents the flux through the pinhole.

This shows that even with a slit width of 50% of the star's FWHM, more than 90% of the available light still enters the spectroscope and means we can get much better spectral resolution with a slit with minimal light loss.

The slit gap should also be matched to the rest of the optics in the spectroscope. A 20 μm gap may be a reasonable "solution" for the star image, but what appears on the CCD? If the collimator and the imaging lens are the same focal length, then the spectral line image will be recorded as 20 μm. With a CCD chip where the pixel size is 9 μm, the image would cover just over two pixels and would be close to satisfying

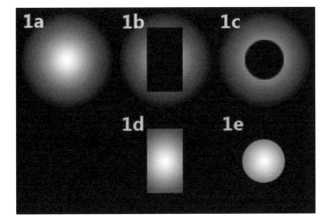

Figure 12.3. Slit transmission efficiency. (CAOS.)

the Nyquist sampling theory. If the pixel size was larger, say 14 μm, we would lose resolution due to under sampling and could just as well use a wider slit – up to 30 μm – without impacting on the resolution.

The ideal slit gap should therefore be assessed on the basis of the telescope plate scale, the seeing conditions, the resolving power of the collimator/imaging lens, and the pixel size of the camera (Also note that if the slit is significantly wider than the star image there will be no edge reflections to guide on).

The length of the slit is not as critical as the width. As the star image is generally so small (<50 μm) a working height of 3 mm is sufficient to allow some freedom in positioning the star image and also allow extended objects (planets, nebulae, etc.) to be imaged. Long slits give rise to distortions in the shape of the observed line. Due to the path length difference of the light from the extreme ends of the slit and the on-axis light, the spectral line forms an arc in which the ends curve toward the shorter wavelengths. The radius of curvature of the arc is approximately the same as the focal length of the imaging lens. This is also seen when prisms are used.

As the lines seen in the spectrum reflect the width of the slit any dust or distortion will also show. These are seen as dark bands or lines along the length of the spectrum, called "transversalium" lines. See Fig. 12.4.

Figure 12.4. Tranversalium lines showing on spectrum due to dirty entrance slit.

Adjustable slits are very useful and can allow initial focusing and alignment with a wide gap. The slit is then adjusted down to the optimum gap for imaging. The adjusting mechanism, which is required to maintain a regular gap over, say, 2 or 3 mm of travel can be complicated. Some designs use V grooves and micrometer heads to move the slit; others have parallel linkages or wedge slides to maintain the gap.

For the amateur there are two options: either invest the time and energy into designing a fully functional entrance slit mechanism, or purchase a commercial slit plate.

Classic Semi-fixed Entrance Slit

The simplest DIY design is based on the steel blades from a pencil sharpener. These are usually very good quality and measure 7.5 mm × 24 mm × 0.6 mm thick with a single ground bevel on the cutting edge. Check the smoothness of the bevel edge by lining up the blades and bringing the edges together. When viewing a bright light the shut off should be complete along the length of the blade. Any minor defects can be removed by rubbing the bevel edge along a thick glass with some 10 μm aluminum oxide to true the edges microscopically. The use of a small fixture will ensure that the edge of the bevel remains square and doesn't get rounded during the polishing.

The bevel slit is mounted on a sturdy frame with a clear aperture of between 3 and 6 mm. The bevel edge should face away from the telescope and towards the collimating lens to reduce unwanted reflections. The front faces can be polished with metal polishing cream to improve the reflection if slit guiding is considered. Again, make sure the bevel edge doesn't get rounded or damaged.

The blades need to be mounted on a solid support frame that will hold them rigidly in place relative to the telescope entrance beam and the collimating lens in the spectroscope. This support, if you are considering guiding from the star reflection on the slit, must be inclined to the optical axis to reflect the image into the guide camera or the guiding transfer lens system. It must also be able to be finely adjusted to bring the image of the slit parallel to the grooves on the grating.

One blade can be epoxy glued across the support frame on the center line and the other lightly clamped in place with a suitable cover plate, the gap adjusted with a small toothpick until it is parallel and as small as required, then the clamp tightened. Recheck that the blade is still parallel. Alternatively both blades can be clamped to give a symmetrical slit. Thin front surface glass mirrors can be used instead of steel blades to give highly reflective slits. See later in this chapter for methods of measuring slit gaps.

Figure 12.5 shows a slit holder used by professional astronomers. It can rotate along a vertical axis that passes through the slit. Thus the reflection from the slit plate can be directed towards the desired direction without changing the slit position. The central part of the holder can rotate in the plane of the reflecting slit plate to change the slit orientation. The reflective slit plate is interchangeable.

Figure 12.5. The slit holder of the TRIPPEL spectrograph at La Palma Observatory. **a**: vertical axis rotation clamps, **b**: rotation clamps, **c**: slit rotation adjuster, **d**: slit jaws adjuster, **e**: focuser rods, **f**: slit plate clamps. (Courtesy D. Kiselman.)

Adjustable Entrance Slits

Using similar blades it is possible by mounting one of them on a sliding base plate to make an adjustable unilateral slit (Ideally the two blades should move together, this gives a bi-lateral symmetrical slit which is always centered on the optical axis). A micrometer head working against a spring controls the position of the moving blade.

Other ideas used by ingenious amateurs include cross linkages at the top and bottom of the blades to maintain a parallel motion, one blade cut at an angle and sliding up and down a guide to open and close the slit. See Fig. 12.6.

Commercial Slits

There are many suppliers of scientific optical slits. The slits available range from laser cut stainless steel slits in widths from 10 μm upwards through etched or engraved glass types to the basic adjustable slit originally used in school spectroscopes. Appendix A in this book lists some current suppliers.

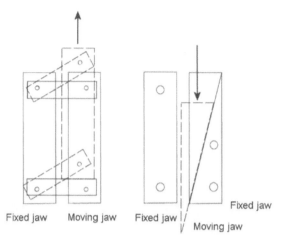

Fixed jaw   Moving jaw   Fixed jaw   Fixed jaw
Moving jaw

Figure 12.6. Designs for DIY adjustable slits.

Setting the Slit Gap

One of the most elegant solutions is to use a laser pointer (see Fig. 12.7). As monochromatic light goes through a very narrow slit, interference patterns are produced. These can be projected onto a suitable wall and the distances between the

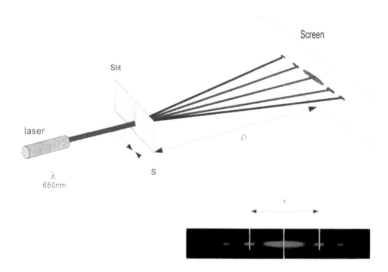

Figure 12.7. Measuring slit width using a laser. (After C. Buil.)

bright fringes measured, there is a direct co-relation between these measures and the actual slit gap. The formula is:

$$S = 3D^*\lambda^*10^{-3}/X$$

where

λ = wavelength of the laser (Å)
S = slit width (μm)
D = distance to the wall or screen (m)
x = distance between the first two bright interference lines (mm)

Using a red penlight laser (6500 Å), distance x will be as shown in Table 12.1 below:

Table 12.1 Slit gap vs. fringe spacing for various screen distances

Slit gap (μm)	Screen distance (D)			
	0.2 m	0.5 m	1.0 m	2.0 m
10	48 mm	96 mm	195 m	395 mm
20	–	48 mm	98 mm	210 mm
30	–	33 mm	66 mm	129 mm
50	–	19 mm	39 mm	80 mm

Projection Method

A small achromatic lens can be used as a projection lens to measure slit gap. A bright light should be used to illuminate the rear of the slit and the lens used to project an image of the slit onto a suitable screen.

Using the thin lens formula:

$$1/S_1 + 1/S_2 = 1/F$$

where F is the focal length of the lens, S_1 the distance to the slit, and S_2 the distance to the screen. The magnification is given by:

$$M = -S_2/S_1 \text{ or } F/(F - S_1)$$

Here's an example, using the above equations. What magnification would we get using a 50 mm focal length lens and a screen at 3 m?

$$1/S_1 + 1/3000 = 1/50$$

$$S1 = 50.85 \, \text{mm}$$

$$M = 3,000/50.85 = \times 59$$

This means a slit width of 25 μm would project onto the screen as an image of 1.48 mm

Other Slit Alternatives

For solar spectrum observing, the reflection from a needle can also be used as a slit. The virtual width of the slit is related to the diameter of the needle and the angular size of the illuminating source.

Suiter, in his *Star testing Astronomical Telescopes,* p. 339, calculates the "glitter" for a sphere as:

$$U = D^* \sin (\theta r/2) \tan \varphi/2 \, 10^3$$

where

U = glitter width (μm)
D = diameter of sphere (mm)
Θr = reflection angle
φ = angle of the source illumination (radians)

Translating this for a cylindrical reflector, we can approximate to:

$$U = D \, \varphi/4$$

For the Sun, with a diameter (φ) of 0.5° (0.0087 radian), this gives a virtual slit approximately 1/300 the diameter of the needle. A 1 mm needle will give an equivalent 4.5 μm slit.

A needle can also be used at the focus of a telescope (or illuminated by an outside light source such as a reference lamp). When applied to an f6 entrance beam, φ = 0.1667 radian, the same needle would give an effective slit of 60 μm. However the reflected beam from the needle is not a "tight" beam and effectively radiates across a wide arc. The use of a secondary focusing lens can improve the emergent light cone.

Photographic Negative for Slit

A photographic negative can be used as a slit. An exposure using fine-grained black and white film (i.e., Ilford PanX Plus) of a high contrast image of a black line on a white background can result in usable slits. The resolution of the film will usually limit this method to slit widths > 25 μm on the negative. This type of slit is obviously not suited for solar work!

The Collimator

The function of the collimator is to accept all the light passing through the entrance slit and produce a parallel beam large enough to illuminate the prism/grating. This means that the focal ratio of the collimator must match that of the telescope. If the focal ratio of the collimator is greater than the telescope, i.e., an f8 collimator on an f6 telescope, then a lot of light will be lost and not get to the grating. It is much better to have an f8 collimator working on an f10 telescope; none of the light is lost, and there is only a small drop in theoretical resolution (see Fig. 12.8). The collimator must be capable of precise focus on the entrance slit to give a parallel output beam. The slit must therefore be aligned with the optical axis of the collimator and be exactly at the focus of the lens.

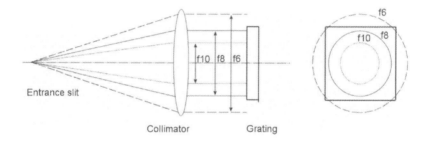

Figure 12.8. Collimator f ratio vs. grating illumination size.

The size of the collimator lens must also be large enough to fully illuminate the width of the grating, i.e., a 30 mm square grating will require a minimum objective diameter of approximately 30 mm. The minimum size will also be determined by the input angle to the grating (See later in this chapter for the effects of grating rotation and anamorphic factors).

The optical quality of the collimator should be good enough to ensure that the projected image of the slit is clean and undistorted. When using narrow slit widths the resolving power (Rayleigh limit) of the collimator must at least match the slit width.

Spherical mirrors (i.e., Ebert, Czerny-Turner designs) can be used as collimators, or lenses (achromatic doublets). The lenses, unlike the mirrors, will obviously give some chromatic aberration. When using cemented doublets (Ex-Binoculars, etc.) look out for the type of bonding material. The more recent UV-cured bonding agents appear to absorb wavelengths below 4,000 Å. A fully anti-reflection coated air-spaced doublet is probably the best lens to use. Camera lenses can be used as collimators, but usually the required f/ratio, aperture, and focal length (approximately 250 mm f6) limit their usefulness.

There are many good suppliers of small mirrors and lenses. A representative list is given in Appendix A in this book.

The Prism as a Dispersion Element

Prisms have been successfully used for over a 100 years as dispersion elements in spectroscopes. They have the benefit of producing a single bright unambiguous spectrum (no second order to overlap!) but at the cost of non-linearity. The dispersion reduces significantly in the red regions, and subsequent spectral analysis requires at least three (or more) reference lines to achieve calibration. See Fig. 12.9.

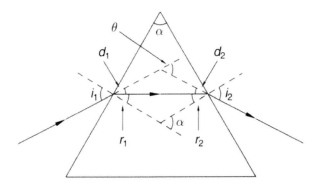

Figure 12.9. Light path through a prism.

The refractive index (n) is measured using Snells Law:

$$n = \sin i / \sin r$$

where í is the angle of incidence to the surface and r is the angle of refraction from the surface.

For a particular wavelength, and hence a particular refractive index (n), the angle of refraction or deviation (D) becomes:

$$D = i - \theta + \sin^{-1}\left\{ n\sin[\theta - \sin^{-1}(\sin i / n)] \right\}$$

At minimum deviation, i.e., when the ray passes through the prism parallel to the base:

$$\text{Angle of incidence (minimum deviation)} = \sin i = n \sin(\theta/2)$$

And the deviation becomes:

$$D = 2\sin[n\sin(\theta/2)] - \theta$$

where θ is the apex angle of the prism. For a $60°$ prism this can be simplified to:

$$D = 2\sin^{-1}(n/2) - 60°$$

Table 12.2 below gives refractive indices for some of the standard glass available:

Table 12.2 Refractive indices for common glasses

Glass type	Wavelength (nm)			
	400	500	600	800
Crown glass (BK7)	1.530	1.521	1.516	1.510
BK4	1.587	1.575	1.545	1.524
Flint glass (F2)	1.652	1.630	1.618	1.608
Dense flint (SF2)	1.684	1.659	1.646	1.635
ED (FPL-53)	1.447	1.441	1.438	1.435
Fused quartz	1.470	1.462	1.457	1.452
Fluorite	1.442	1.436	1.431	1.430
Rock salt	1.570	1.551	1.544	1.537

BK7 prisms can readily be found in most optical catalogs. A better solution is to use a flint F2 or SF2 $60°$ prism, which has a higher dispersion (approximately 10% more).

To understand the spectrum produced by a prism, using the data in Table 12.2, we'll consider blue/UV light at 400 nm and red light at 800 nm using a 30 mm, $60°$ F2 prism, and a center wavelength of 500 nm:

Angle of incidence (minimum deviation) $= 1.63\sin(60/2) = 0.815, i = 54.5°$

Deviation angle for 500 nm $= 2i - \theta = (2 * 54.5°) - 60° = 49°$

Dispersion spread, $\Delta D = (2\sin(\theta/2)/\cos i)\Delta n = (2\sin(30°)/\cos 54.5°)(1.652 - 1.608)$

$$\Delta D = 0.0758 \text{ radian} = 4.35°$$

This means that the spectrum between 400 and 800 nm is spread over a beam 4.35° wide.

If we now consider imaging this spectrum, say, with a 100 mm fl lens (F):

Length of spectrum at the focal plane $= F * \Delta D = 100 * 0.0758 = 7.6$ mm

This gives an average dispersion/plate scale of $400/7.6 = 52.6$ μm/mm (526 Å/mm).

A more accurate solution can be found by using the Hartmann dispersion formula:

$$n(\lambda) = A + B/(\lambda - C)$$

where the constants A, B, and C are given by:

$$C = \frac{\frac{n_1 - n_2}{n_2 - n_3} \lambda_1 (\lambda_2 - \lambda_3) - \lambda_3(\lambda_1 - \lambda_2)}{\frac{n_1 - n_2}{n_2 - n_3}(\lambda_2 - \lambda_3) - (\lambda_1 - \lambda_2)}$$

$$B = \frac{n_1 - n_2}{\frac{1}{\lambda_1 - C} - \frac{1}{\lambda_2 - C}}$$

$$A = n_1 - \frac{B}{\lambda_1 - C}$$

Some typical values of these constants (based on λ measured in nm) are given in Table 12.3.

Table 12.3 Hartmann constants

	A	B (10^{-8})	C (10^{-8})
Crown glass	1.500	0.35	−2.5
Dense flint	1.650	0.21	1.5
Fluorite	1.429	0.53	3.6

Spectral lines produced by a prism are usually curved due to the height of the slit, causing the off-axis beam from the ends of the slit to enter the prism obliquely. The path through the prism is effectively longer for these rays, so the deviation is greater. The image of the slit then becomes an arc, with the ends curved towards the shorter wavelengths. The curvature of this arc also changes with wavelength due to the differing refractive indices. To minimize this effect short slit gaps are preferred.

Resolution

The resolution of the prism spectrum is generally dependent on the star image size or the slit width (if used), the base length of the prism, the focal length of the camera lens, and the pixel size of the CCD.

Let us assume an ideal entrance slit and a pixel size of 6 μm in the above example:
Minimum CCD resolution (Nyquist) = 12 μm
This then gives 526 * 0.012 = 6.3 Å

(A better and more accurate result would be obtained by analyzing the FWHM results from a reference lamp emission line.)

To improve the spectral resolution from a prism, multiple prisms can be used to increase the effective base length, each prism set at the minimum deviation angle to the previous. Early spectroscopes used up to six or even eight prisms in the "train" to get maximum resolution.

The theoretical resolution of a prism was given earlier:

$$R = \Delta n / \Delta \lambda * \text{base length of prism(s)}$$

where $\Delta n / \Delta \lambda$ is called the dispersion of the prism.

For common glass materials (n in the range 1.4–1.6), an approximation (within 4% – exact when n = 2) is:

$$\Delta \theta / \Delta \lambda = n \, \Delta n / \Delta \lambda$$

A first approximation may be achieved by:

$$\Delta n / \Delta \lambda = (1.630 - 1.618)/(600 - 500) * 100,000 = 120 \, \text{mm}^{-1}$$

For a prism base length of 30 mm, this gives:

$$R = 3600$$

i.e., a theoretical resolution of just less than 0.2 nm (2 Å) in the green.

The following Table 12.4 gives some dispersion values (expressed in change per μm) for various materials.

Table 12.4 Dispersion values for crown and flint glass

	400 nm	600 nm
Crown glass	0.013	0.0036
Flint glass	0.051	0.010

Abbe Numbers

You may see glass specified by its Abbe number. This is a measure of the refraction index of the glass and used to define the type of glass (i.e., crown/flint/dense flint, etc.)

$$V = (n_D - 1)/(n_F - n_C)$$

where n_D is the refractive index at the Fraunhofer line D (589.2 nm), n_F (486.1 nm) and n_C (656.3 nm).

Glasses with an Abbe number below 50 are classified as flint glasses and above 50, as crown.

Gratings as a Dispersion Element

The most popular dispersion element available to the amateur today is the grating. Earlier we gave an overview of grating theory and performance. Either transmission or reflection gratings can be used in the spectroscope.

Gratings come in various lines per millimeter (l/mm); from 150 to 2400 lines. The 2400 l/mm grating has the potential to give very good resolution but in doing so will also disperse the visible spectrum over a far longer image. This means that not all the spectrum will fit in the field of the camera, and multiple exposures may be needed to capture the whole picture. A 1200 l/mm grating and a MX7C CCD camera, for example, will need about twelve images to cover the 1st order solar spectrum.

A good starting point is to use a blazed 300, 600, or 1200 l/mm grating. This gives a good compromise between plate scale, resolution, and spectrum brightness. Buy the largest grating your budget will allow; this is the "heart" of the spectroscope. Even if your initial design doesn't fully illuminate a large grating, further "upgrades" can take advantage of it. Unless you have a need to do spectral analysis in a specific region (i.e., UV or near IR), it's best to go for the standard "visual" blazing angle of 550 nm.

Blazed Gratings

The spectrum formed from a reflection grating is reflected from the numerous facets cut in the material. Maximum efficiency is achieved when the light is directly reflected off the faces of the facets. In a non-blazed grating this condition occurs in the zero order image. By altering the face angle of the facet (γ) it is possible to preferentially deviate the maximum amount of light into one of the first order spectra (Fig. 12.10). This can dramatically improve the efficiency.

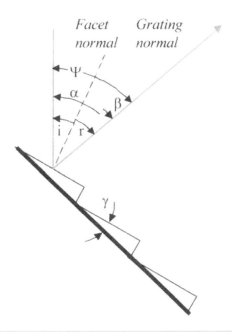

Figure 12.10. Details of blazed reflection grating. (J. Allington-Smith.)

At the blaze wavelength λ_B

$$\alpha + \beta = 2\gamma$$

the grating equation

$$nN\lambda = \sin\alpha + \sin\beta$$

becomes:

$$n\,N\lambda_B = 2\sin\gamma\cos(\psi/2)$$

where $\Psi = \alpha - \beta$.

Note that a grating blazed for 550 nm in the first order will also give an increased efficiency to the same wavelength in the second order.

All blazed gratings have an arrow printed or marked on one of the edges; this indicates the direction of the blazed first order spectrum.

In the optical layout of the spectroscope you will need to define the angle at which the grating will be set relative to the collimator (α) and the imaging lens (β) to maximize efficiency and bring the anamorphic factor close to unity (See later).

For classical designs this angle (Ψ) is about 38°. Generally the efficiency falls as the angle between the collimator and the imaging lens increases.

As mentioned before, the reflection grating is very delicate and should be protected at all times. Wear cotton gloves when handling to prevent fingerprints. The mounting frame should hold the grating (without undue pressure) firmly in a square and vertical condition. If designed to rotate, the shaft should preferably be mounted in a precision ball bearing, to give smooth and consistent movement.

The design of the spectroscope should allow the grating to be positioned as close to the collimator as possible, and if full spectral coverage is required, the grating should be mounted to allow rotation of at least 20°. The center of rotation should be through the center of the grating and aligned with the front surface of the grating. To accurately position the grating a micrometer head/sine bar can be used to rotate the fixture.

Anamorphic Factor

You need to consider how the collimating and camera lens will be aligned with the grating. Due to the limited dispersion of the spectrum it is usual to mount the grating on a rotating platform to bring different parts of the spectrum into the field of view of the camera. Use the grating equation:

$$nN\lambda = \sin\alpha + \sin\beta$$

The angle of the incoming (α) and reflected (β) (or transmitted) light beam to the grating must be considered (Fig. 12.11). If the grating (width, W) acted like a normal mirror surface the input beam from the collimator (d_1) and the output beam to the camera (d_2) would be the same circular shape (as is the case with the zero order). When the grating is tilted, the input width (W_1) will be:

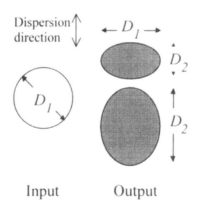

Input   Output

Figure 12.11. The namorphic factor, showing the distortion of a reflected beam from a grating. (J. Allington-Smith.)

$$W_1 = W^* \cos\alpha$$

and the outgoing beam width (W_2) to the camera:

$$W_2 = W^* \cos\beta$$

The input and output beam diameters will depend on the size of the collimator objective. If this is sized to fully illuminate the grating width, then D_1 should be the same as or slightly larger than the effective grating width:

Input beam diameter, $D_1 \geq W1$
Output beam diameter, $D_2 \geq W2$
The anamorphic factor (A) is the ratio between the two:

$$A = d_2/d_1 = \cos\beta/\cos\alpha$$

The effective magnification in the dispersion direction then changes by this factor. As the grating tilt angle changes so does the anamorphic factor, which can lead to a variation of up to 4% in the plate scale across the spectrum; the number of lines illuminated will also vary in proportion to $1/\cos\beta$.

To minimize the effects of the anamorphic factor, it should be designed as close to unity as possible. This is achieved in the Littrow design.

With blazed gratings we can position the grating such that we have *normal to collimator* or *normal to imaging lens* (see Fig. 12.12).

The normal to imaging lens configuration is usually the preferred layout.

Details of suggested grating mounting plates are given later, when discussing the construction of the spectroscope.

The Imaging Lens

The aperture of the imaging lens must be sufficient to collect all the diffracted light from the grating and be positioned as close as possible to the grating.

The image of the slit produced on the camera/CCD is:

$$Is = S^* F_2/F_1$$

where S is the width of the slit (μm), F_1 the focal length of the collimator, and F_2 the focal length of the camera. The ratio F_2/F_1 is the magnification factor. See Fig. 12.1.

This means that a 50 μm entrance slit can be projected as a 25 μm image by choosing a collimating lens with twice the focal length of the camera lens, i.e., 200 mm vs. 100 mm. It also reduces the effective linear dispersion, giving brighter, shorter spectra. Depending on the pixel size of the CCD this can also improve the sampling ratio. In the case of the Littrow design the collimating lens doubles as the imaging lens, so there is no magnification factor.

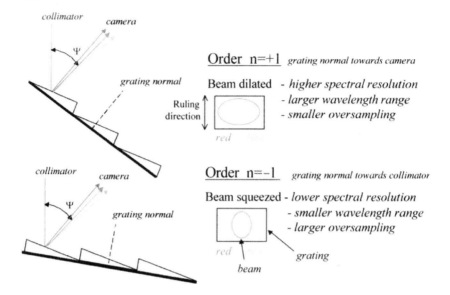

Figure 12.12. The effect of grating normal to camera vs. grating normal to collimator. (J. Allington-Smith.)

The imaging lens should also be capable of focusing the whole spectrum onto the CCD, bearing in mind that the chromatic aberration in achromatic doublets will bring the red region to a different focal point than to the blue.

The linear dispersion or plate scale is:

$$D = \cos\beta \; 10^7/n \; NF_2 \text{Å/mm}$$

The optical axis of the spectroscope should lie on the center of the slit, through the center of the grating to the center of the CCD, with all optical surfaces (collimator and imaging lens square to the axis).

In a classical spectroscope design, most amateurs use old camera lenses of 50–200 mm focal length. By using standard T ring adaptors it's easy to interface with CCD cameras, Webcams, and DSLR's. This lens is usually set fully open (f1.8–f2.8) and focused at infinity.

The Eyepiece/Camera/CCD

When imaging the spectrum, the pixel size, quantum efficiency (QE), and overall size of the CCD chip are important. The smaller the pixel the better the resolution (subject to effective sampling); the better the QE the better the camera response to various wavelengths, giving lower exposure times, and the larger the chip the more of the spectrum recorded on each exposure.

If the spectroscope is being used on stars, just bear in mind the typical exposures (for a 5 mag star) are in the order of 6 * 5 min = 30 min total. The solar spectrum only needs 1/20–1 sec exposure.

The dark field exposures and flat field exposures are similar to those in normal astro-imaging.

Attaching and Focusing the Camera/CCD

Most CCD cameras have a back focus requirement (distance from front flange to CCD chip surface) of about 10–12.5 mm. A DSLR normally requires 45 mm (55 mm if you include a T2 adaptor). A standard camera lens focuses on infinity at a back focus distance of 45 mm from the rear flange, but the inner element of the lens may be further back and reduce the "clearance" to less than 22 mm. The normal length of a 1.25″ T thread nosepiece is around 30 mm. All these variations need to be taken into account when designing the camera attachment.

If a mirror or an achromatic doublet is used as the imaging lens, then a standard helical focuser and spacer tubes can be used to advantage. Using camera lenses requires an adaptor to accommodate the CCD. A typical solution is to use a modified "camera macro extension tube." An adaptor plate (see Fig. 12.13) can be machined to allow 1.25″ fittings to be used, or T thread adaptors.

Whatever method is used, it must hold the camera/CCD square and secure and still allow some rotation to align the spectrum with the horizontal axis, and if using a doublet, a means of adjusting the focus for the blue and red extremes.

Spectroscope Design Summary

Based on the above design rules we can now look at a checklist for the design of a spectroscope. For ease of understanding it will be assumed we want to build a classical design with a rotating grating (A Littrow design is obviously similar but with the collimator also acting as the imaging lens).

Questions that need to be asked include:

1. How will the spectroscope be attached to the telescope? T thread adaptor? 2" nosepiece? Special adaptor?
2. Do you want to include a slit or provision for a future slit? A reflective slit for guiding? If guiding, what camera? Mounting requirements? Focusing the guide camera?
3. What slit gap? Or multiple slits? Adjustable slit? Do you need back illumination?
4. What is the focal ratio of the collimator? Based on the available telescope? Other telescopes you may use?
5. What target for R value? What size of grating? How many l/mm? Interchangeable gratings?

Figure 12.13. Camera lens adaptor.

6. What angle between the collimator and imaging lens?
7. What focal length imaging lens? Magnification factor? Interchangeable?
8. Which camera(s) will you use? How big is the CCD chip? What's the size of the pixel? How will it be attached to the spectroscope? Focusing the camera?
9. Is weight important? Physical constraints? Should a fiber optics layout be considered?

There are no definitive answers to these questions. Each spectroscope is different, but it must be clear from the outset what design decisions are important to the user. It's very difficult to change a design halfway through; you usually end up starting again from scratch.

The key inputs for you are:

Resolution required	R value
Size of grating	W
Telescope focal ratio	fr_t
Imaging lens focal length	F_2
Angle between the collimator and imaging lens	Ψ
Camera pixel size	p

Note: To make the calculations easier it is usual to assume a starting point for the grating l/mm. For low resolutions, R< 2000, start with a 300 l/mm option; for R>10000 a 1200 l/mm would be a good starting point.

Let's design a spectroscope based on:

Resolution required	R>1000, 5.5 Å in the 1st order green spectrum
Size of grating	30 mm × 30 mm
Telescope focal ratio	f6
Imaging lens focal length	135 mm f2.8 camera lens
Angle between the collimator and imaging lens (Ψ)	38°
Camera pixel size and number	9 μm, 765 (ST-7)
Chip size (dispersion direction)	6.9 mm

Assumptions

- The center wavelength (n_0) chosen is green 550 nm.
- For the first approximation we'll consider a 300 l/mm grating (If this does not provide the necessary resolution, we can recalculate the other parameters based on a 600 or 1200 l/mm grating).

Calculations

Note: The spreadsheet SimTrans V3 (See Appendix D) is an ideal resource for grating spectroscope design. When the above parameters are entered, the outputs calculated give all the data necessary to complete the design. See Fig. 12.14.

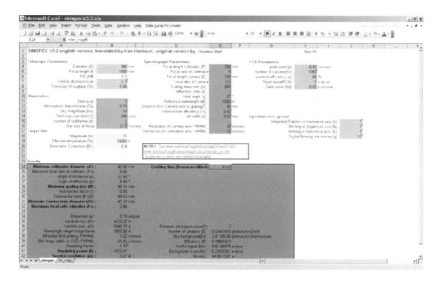

Figure 12.14. SimSpec data entry screen.

The design calculations can also be carried out by following the sequence detailed below:

Consider the incident (α) and diffracted angles (β) at the grating.

From the grating equation, using first order (n = 1) and $\Psi = \alpha-\beta$:

$$nN\lambda_0 = \sin\alpha + \sin\beta$$

$$\sin\alpha + \sin(\sin\alpha - \psi) = n\,N\lambda_0$$

$$\mathrm{Sin}(\alpha - \psi/2) = nN\lambda_0/2\,\mathrm{Cos}(\psi/2)$$

$$\mathrm{Sin}(\alpha - \psi/2) = 1{}^{*}300{}^{*}550{}^{*}10 - 6/2\cos(38°/2)$$

$$\mathrm{Sin}(\alpha - 19) = 0.08725$$

$$(\alpha - 19) = 5°, \alpha = 24° \text{ and hence } \beta = -14°$$

(This means that the diffracted ray lies on the opposite side of the grating normal.) The anamorphic factor (A) would be 1.0621.

Based on fully illuminating the width of the 30 mm × 30 mm grating, the input beam (d_1) should be:

$$\text{input beam diameter, } D1 = W^* \cos\alpha$$

$$= 30^* \cos 24°$$

$$= 27.5\,\text{mm}$$

The collimator should therefore have a minimum diameter of 27.5 mm.

Let's now look at the theoretical linear spectral image resolution. With a 9 μm pixel the Nyquist sampling tells us the resolved image size will be 18–20 μm; let's work with 20 μm. If this is to be representative of the resolution required (5.5 Å), then the dispersion of the grating needs to be 0.275 or 275 Å/mm.

From the equation earlier in this chapter:

$$\text{The linear dispersion or plate scale (D)} = \cos\beta 10^7/n\,NF_2 \text{ Å/mm}$$

Substituting:

$$D = \cos\beta 10^7/1^*135^*300$$

$$D = \cos(-14°)10^7/40.500$$

$$D = 239 \text{ Å/mm}$$

The standard 300 l/mm grating gives a plate scale of 239 Å/mm and 2.15 Å/pixel, giving a theoretical resolution of 4.78 Å, or R = 1150, which exceeds the requirements.

The exit beam from the grating will be elliptical in shape, due to the anamorphic factor (A) and requires an imaging lens with a minimum diameter:

$$d_2 = d_1^* A = 27.5^*1.0621 = 29.2\,\text{mm}$$

Does the 135 mm camera lens meet this requirement? At f2.8 the clear aperture is 48.2 mm, which is greater than the minimum. We can also consider the full illumination of the CCD chip.

To achieve maximum transmission efficiency (etendue)

$$X_1 + X_2 = F_1$$

where:

X_1 is the distance from the collimating lens to the grating,
X_2 the distance from the imaging lens to the grating, and
F_1 is the focal length of the collimator.

The distance between the grating and the imaging lens (X_2) should be kept as short as possible, consistent with the mechanical constraints of locating the lenses, etc. A target value of 50–60 mm should be considered. Let's assume 60 mm.

The width of the CCD chip is 9 μm * 765 = 6.9 mm. The necessary correction would be:

$$= X_2{}^*\text{Chip size}/F_2$$

$$= 60{}^*6.9/135$$

$$= 3.07 \text{ mm to be added to the diameter of the imaging lens}$$

This brings the minimum unvignetted imaging lens diameter to (29.2+3.07=32.27 mm), which is still less than the 135 mm f2.8 camera lens aperture.

To match the telescope focal ratio (f6), the collimator focal length must be $d_1 * f_1 = 27.5 * 6 = 165$ mm and be positioned close to (165–60) = 105 mm from the grating. A 30 mm diameter f5.5 achromatic doublet would probably be a reasonable compromise (It could be stopped down to 27.5 mm to give f6).

The slit gap will be: (F_1/F_2) * CCD image size = (165/135) * 20 μm = 24.5 μm

So here we have a first design:

Slit	24.5 μm, say 25 μm
Collimator	30 mm f5.5 (stopped down to 27.5 mm f6)
Distance to grating	105 mm
Total angle incident diffraction beam	38°
Grating	30 mm × 30 mm, 300 l/mm
Distance to imaging lens	60 mm
Imaging lens	135 mm f2.8 camera lens
Camera	9 μm pixel × 765 pixel wide
Plate scale	239 Å/mm, 2.15 Å/pixel
Theoretical resolution	4.78 Å, or R=1,150

Reference Lamps

A reference light source is required to do any serious calibration of your spectra. It's also preferable to have the reference light mounted on the spectroscope to allow a calibration spectrum to be taken before and after the target spectrum. This can be achieved by having a small flip mirror (in front of the slit) that can illuminate the slit or by using a small prism to direct the light into the bottom of the slit. In older professional spectroscopes they used a Dekker mask (Decker in the United States) (see Fig. 12.15).

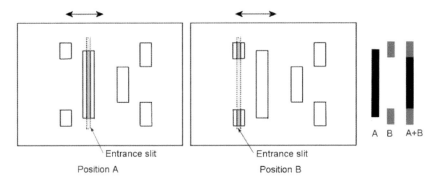

Entrance slit   Entrance slit
Position A   Position B

Figure 12.15. Dekker mask, showing target spectrum in the center (Position **a**) with reference spectra above and below (Position **b**).

When slid into place this mask would cover the working length of the slit and provide apertures immediately above and below the target spectrum for comparison. Fiber optics can also be used to present the reference light to the slit.

In some spectroscope designs the shape of the recorded spectrum can be distorted due to the aberrations in the optics. This complicates the relationship of the reference spectrum to the target spectrum.

The standard reference lamp used by most amateurs is a neon light bulb. This has a very well-documented spectrum and covers wavelengths of between 580 and 700 nm (Additional faint lines can be found in very long exposures, down to 400 nm). These are readily obtained from electronics stores, as neon indicator lamps. The operating voltage is usually around 80 V (a suitable resistor in series will drop the mains voltage to the required 80 V). It is possible to construct a low voltage inverter to allow them to be used from a 6 or 12 V supply. This removes the risk of using 110 or 240 V electrical connections at the spectroscope. Figure 12.16 shows a typical circuit diagram (for personal use only and not for commercial use).

It is also possible to construct a low voltage driver circuit for the neon by re-using the PCB removed from a 12 V fluorescent work lamp. See Fig. 12.17.

Low voltage fluorescent lamps themselves can also be used as reference lights, the mercury line at 480 nm complementing the neon.

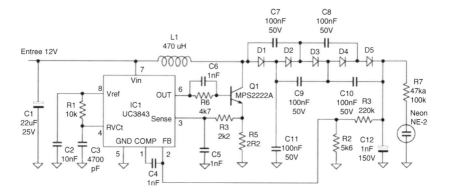

Figure 12.16. Circuit diagram for neon lamp invertor – non commercial use only. (Shelyak.)

Figure 12.17. A 12 V fluorescent lamp driver circuit.

Testing and Calibrating Spectroscopes

During construction and assembly of the spectroscope it is essential that the various optics sit square to the optical axis at the correct design spacing. Care should be taken at all times to protect the delicate grating from dust and damage.

The optical alignment can be checked and verified using a laser collimator.

Laser Collimator

A small red laser pointer can be modified by removing the nosepiece and adding a mask made from a "precision" pinhole in aluminum foil. This is necessary to improve the original "optical axis" of the laser beam, which can be well off-center and spread to an unacceptably large image.

The pointer was mounted, using six small screws, at 120°, into a surplus 1.25″ × 1″ extension tube (Fig. 12.18). With trial and error the pointer was accurately positioned in the center of the tube and aligned with the axis such that there was no movement of the spot when projected on a wall 4 m away and the tube rotated. This is much cheaper (and more accurate!) than some of the commercial models.

Figure 12.18. Homemade laser collimator.

By temporarily masking the slit gap, leaving a 1 mm gap in the center of the slit (to define the position of the optical axis), the laser can be set up square in front of the slit and the beam used to check that the collimator and grating holder are central to the optical axis. A 1.25″ Barlow with the lens and chrome portion removed (i.e., a 1.25″ sleeve) can be used to get the collimator central.

Later, when the grating is positioned on its mounting frame and set to the zero order position, the laser can also be used to confirm the alignment of the camera lens. As mentioned earlier, the laser can also be used to precisely measure the slit gap.

Testing the Collimator Alignment and Focus

Once all the components have been assembled we need to check out and test that it works!

By providing a suitable viewing port in the body of the spectroscope (a 25 mm diameter hole centered on the optical axis behind the grating holder is sufficient) the collimator and slit can be quickly and easily checked for final alignment. Remove the grating for safety and use a small finder scope (30 × 6 or similar focused on infinity) positioned behind the hole to look directly into the collimator. The slit gap should be seen in clear and tight focus. Any out of focus can be corrected by very fine adjustment of the collimator (or slit position).

Optical Testing

This can be done on the bench using a normal light bulb or better still an energy-saving flourescent bulb. Position the spectroscope so that the front of the slit is fully illuminated by the light source. If the slit is adjustable, set the slit to give a visible slit gap (around 30 μm). If you can, fit a low power eyepiece (25 mm focal length) at the focus of the imaging lens. Now rotate the grating to bring the bright zero order image into the middle of the field, and focus. You should see a clear, tight, straight line (the image of the slit). If there is any distortion or one edge is in focus and the other out of focus this indicates something is out of alignment.

You should do the same check with your camera in position and use a series of short exposures. As you come towards focus you'll see the image of the slit get narrower until it's just a bright line with sharp edges. Mark the camera focus adaptor. Make a note of the micrometer reading (if fitted) or mark the adjusting screw so you can repeatedly move the grating and return to the zero order image.

If the zero image doesn't appear vertical (or horizontal) in the frame, rotate the camera body until, when measured, there is no difference in the pixel column (or row) between the top of the image and the bottom. Mark this position on the camera adaptor. Remove the camera, and re-connect to the focus and registration mark; the camera should show the zero image well focused and in the correct orientation.

Once you achieve a good zero image, slowly rotate the grating until a spectrum comes into view. Note the brightness. Now rotate the grating back in the opposite direction until once again you see a spectrum. Which one was the brighter? The first or second? The brighter spectrum is the blazed first order; this is the one you want.

Make a note of the direction of rotation of the grating.

Set the spectroscope grating to give a section of the first order spectrum and cover the body of the spectroscope with a heavy dark cloth. Take an overexposed image of the spectrum (3 or 4 sec). Now remove the cloth and shine a bright torch onto the joints of the spectroscope box; at the same time take an exposure similar to the first. What we are trying to determine is whether or not the unit is light tight. Any flaring or brightening of the image indicates light is getting into the spectroscope. You can apply Gaffa tape around the joints if the sealing is poor. Any residual stray light will affect the performance of the spectroscope.

If you are using an adjustable slit, set the grating back to the zero order position and ensure the focus and orientation is correct. Now slowly close the slit gap; take another exposure and see if the gap still appears regular, with no edge distortions or signs of taper. Keep slowly closing the gap and taking images (You may have to increase the exposure time as the intensity of the image reduces with a narrowing gap). When the slit is almost completely closed, only a couple of pixels wide on the image frame, does it still look regular? This is close to the "normal" operating position for the grating for maximum resolution. Mark the adjusting screw on the slit mechanism. We need to be able to repeatedly return to this position.

Any roughness of the slit edge will show during this test. If you're lucky it may only be some dust on the slit. This can be removed by opening the slit a little and gently wiping along the length of the slit gap with a soft wooden toothpick. If the slit jaws themselves are damaged or rough, then this means you need to rework the edges.

The obvious candidate for testing your new spectroscope is the Sun. It's easy to get a usable spectrum without even mounting the spectroscope in a telescope. Just point it so that the Sun shines directly on the slit (even a bright sky can give results). The slit should be reasonably open, i.e., a visible gap. By turning the grating you should quickly find the first order spectrum. Magnificent view!!

CAUTION: NEVER LOOK DIRECTLY AT THE SUN WITH ANY OPTICAL AID

The Fraunhofer lines should be visible. There are a couple of easy targets – three lines in the green (magnesium) and two in the yellow (sodium). If the slit is aligned with the grating and the camera lens is focused on the eyepiece/CCD these lines should be clearly visible. If not, try rotating the slit slightly; it may also help to narrow the slit gap. Check that the grating rotation is smooth and allows coverage of the complete spectrum; also check that focus can be reached in the blue and red regions of the spectrum and mark the focuser accordingly.

Calibration

By now we should have a functional spectroscope. We know the dispersion, plate scale, and theoretical resolution. All that is left to do is to calibrate the slit adjustment (if adjustable), draw up a calibration curve for the grating micrometer, and confirm the actual resolution.

The slit gap can easily be measured using the procedure detailed earlier in this chapter. A suitable scale should be attached to the spectroscope to allow the slit gap to be repeatedly dialed in.

Using either the Fraunhofer lines in the solar spectrum, or a reference light source (fluorescent, neon, etc.) use the micrometer head to rotate the grating, bringing a known spectral line to the middle of the frame; record the micrometer reading. Do this for as many lines as needed to plot a curve of micrometer reading to wavelength (Fig. 12.19). Repeat the measurements a few times to ensure the repeatability. This will allow you to use the micrometer to set any central wavelength in the frame and also tell you what change in reading you'll need to index frames when building up a complete spectrum.

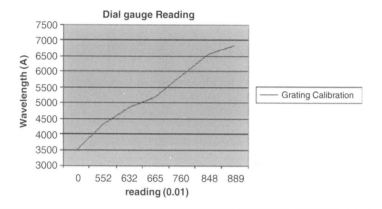

Figure 12.19. Sample calibration curve for dial micrometer.

Using a reference light, say neon, we can now determine the actual resolution of our spectroscope. Set up the spectroscope with the neon light shining onto the slit. Rotate the grating to bring one of the neon emission lines to the center of the frame. Note the wavelength of this line. Now, slowly close the slit gap until there's no reduction in the width of the observed line. Check the focus of the camera. Take a few exposures at varying durations; at this stage we're trying to find the minimum exposure to give the clearest image of the line. Once we have the best image, import the frame into VSpec and calibrate it to the reference lines.

We can either use the built in FWHM function or measure the line shape manually. To do it manually, cut and expand the section of the spectrum containing the reference line; note the peak intensity of the line, divide this by 2, and then move the cursor across the line at this Intensity value. Note the wavelength reading at the front edge and at the rear. The difference between the two readings gives a measure of the line width at half max. Repeat this for other lines to get an average value. The resolution, R value, is $\lambda/\Delta\lambda$. Divide the wavelength of the reference line by the FWHM value found. The plate scale (Å/pixel) can be used to confirm the number of pixels covered by the FWHM image (This provides a double check that we are working within the Nyquist sampling theory). This is then the resolution of your spectroscope.

If the grating, imaging camera lens, or CCD camera is changed, the spectroscope should be recalibrated.

Web Pages

http://spectroscopy.wordpress.com/2009/05/22/slitpinhole-flux-calculator/#more-525
http://www.astroman.fsnet.co.uk/needle1.htm
http://www.cfai.dur.ac.uk/old/projects/dispersion/grating_spectroscopy_theory.pdf
http://www.solarphysics.kva.se/LaPalma/spectrograph/spectrograph.html
http://articles.adsabs.harvard.edu//full/1895ApJ.....1...52W/0000052.000.html
http://articles.adsabs.harvard.edu//full/1995ASPC...71...18S/0000018.000.html
http://www.astr.ua.edu/keel/techniques/spectra.html

CHAPTER THIRTEEN

Prism Spectroscope Designs

The first astronomical application of the spectroscope was to use a prism in front of the telescope objective. This allowed early scientists to develop their theories of star classifications from the observed spectra. The last chapter gave details of how to calculate the deviation and dispersion of prisms.

Objective Prisms

Objective prism spectroscopes are the easiest to construct. Any suitable prism can be securely mounted at its deviation angle in front of a camera lens or telescope objective. This slit-less design gives wide field coverage and bright spectra. If the prism is smaller than the objective a masking screen needs to be placed around it to prevent stray light getting into the telescope. Consider the focal length of the telescope, the dispersion of the prism, and the size of the CCD.

Some method of rotating the prism should be found; it allows the spectrum to be rotated in the field of view, which prevents or at least minimizes interference from other field spectra.

When used on camera lenses, the prism can be mounted in an old lens hood, fitted to a suitable plate, or inserted into a filter holder (i.e., Cokin series). This allows rotation. See Figs. 13.1 and 13.2.

Sometimes large objective prisms become available through secondhand sites such as Astromart. In the picture below (Fig. 13.3), a 4° 100 mm diameter BK4 prism was found and mounted in a 100 mm PVC plumbing connector. This was then used on an ED80 refractor.

K.M. Harrison, *Astronomical Spectroscopy for Amateurs*, Patrick Moore's Practical
Astronomy Series, DOI 10.1007/978-1-4419-7239-2_13,
© Springer Science+Business Media, LLC 2011

Figure 13.1. A 20° objective prism mounted in a Cokin filter adaptor.

Figure 13.2. Objective prism arrangement. (R. Hill.)

Figure 13.3. A 4° objective prism on an ED80.

Other Prism Spectroscopes

Prisms can be used as a dispersion element to construct good, functional spectroscopes. They have the advantage, as mentioned earlier, of providing a bright, unambiguous spectrum. Chapter 12 details the optical and dispersion characteristics of prisms.

Prisms in Collimated Beams

Prisms can be mounted behind an eyepiece (see Fig. 13.4), the exit pupil forming a collimated beam. The spectrum can then be imaged with a standard camera lens. An afocal camera adaptor allows the lens to be tilted to the deviation angle. The dispersion/plate scale can be calculated, and a Barlow lens, positioned to give a collimated beam, can also be used.

The Besos design by CAOS engineers is an interesting variation. It includes a slit, a guiding system, and a 60° flint prism. See Fig. 13.6.

An Amici prism can also be used. It has the benefit of giving an undeviated spectrum, which makes the mounting arrangement more straightforward. A prime focus camera adaptor tube can be used to hold the Amici prism in place behind an eyepiece (or Barlow). See Fig. 13.5.

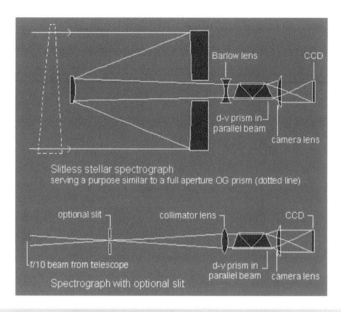

Figure 13.4. Amici optical layout. (M. Gavin.)

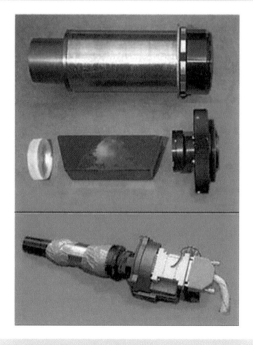

Figure 13.5. Amici prism arrangement. (M. Gavin.)

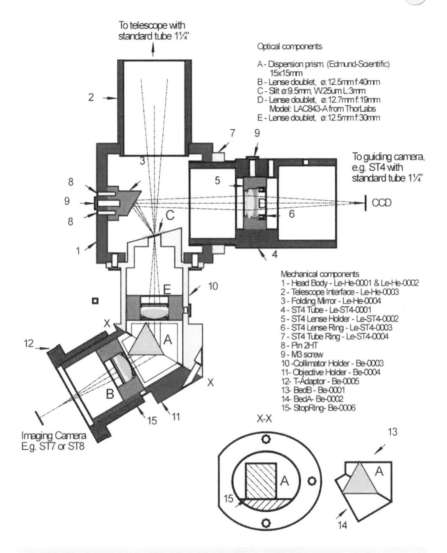

Figure 13.6. Besos prism spectroscope – optical layout. (CAOS.)

Traditional Prism Spectroscope

The laboratory design of a classical prism spectroscope is a collimator (with slit) rotating prism table and an imaging telescope. The early astronomical spectroscopes basically replicated this design (see Fig. 13.7).

Figure 13.7. Standard school spectroscope layout showing the slit and collimator on the LHS, prism table center, and imaging telescope on the RHS.

The main design differences are that the astronomical type must be capable of interfacing to and moving with the telescope and that the collimator focal ratio must match that of the entry beam. A solid support frame holding the slit, collimator, and prism, which has an adaptor to fit the focuser on the telescope, is mandatory. The size of the collimator lens is restricted by the size of the prism. All the available light from the slit should be capable of passing through the prism. The prism must be square to the collimator but set at the minimum deviation angle.

The imaging lens can be mounted on a swivel arm, the center of the pivot in the center of the prism and a micrometer used to position the lens in the diffracted spectrum. Usually a camera lens of focal length similar to or smaller than the focal length of the collimator is used, focused on infinity.

The calibration of the micrometer (using reference lamps or the Sun) is critical due to the non linear nature of the prism spectrum. A reference lamp can be used to illuminate the rear surface of the prism to give wavelength calibration. See Fig. 13.8.

Classic Littrow Prism Spectroscope

A back silvered 30° prism can be used to create a classical Littrow spectroscope. Positioned in front of a suitable collimating lens and tilted to the minimum deviation angle it can give the same dispersion as a 60° prism. See later for typical Littrow construction details.

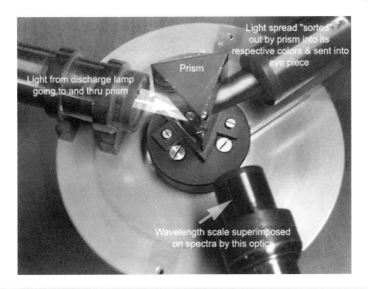

Figure 13.8. Traditional prism spectroscope. (WIKI.)

Efficiency of a Prism Spectroscope

Single-prism systems produce a bright spectrum, with no light lost to a zero order image. Larger prisms and the use of multiple prisms reduce this benefit, due to the additional light losses both from the surfaces and the thickness of the glass.

Transmission efficiencies of more than 90% can be achieved.

Further Reading

Harrison, G.R., Lord, R.C., Loofbourow, J.R. *Practical Spectroscopy.* Prentice Hall Inc (1948).

Houk, N., Newberry, M.V. *A Second Atlas of Objective Prism Spectra.* University of Michigan, Ann Arbor (1984).

Martinez, P. (Ed.). *The Observers Guide to Astronomy*, Vol. 2. Cambridge University Press (1994).

Thorne Baker, T. *The Spectroscope.* Second Edition. Bailliere, Tindall and Cox (1923).

Web Pages

http://www.lpl.arizona.edu/~rhill/spect/spect.html

http://www.iop.org/EJ/article/1538-3881/134/3/1072/205206.text.html

CHAPTER FOURTEEN

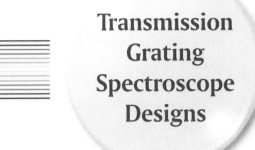

Transmission Grating Spectroscope Designs

Most amateurs will start spectroscopy with the use of filter transmission gratings. They can be used as objective gratings or with the telescope. The most popular method is in the converging beam. This basic arrangement was described earlier. It is also possible to mount transmission gratings into a collimated spectroscope configuration, with or without an entrance slit.

Objective Gratings

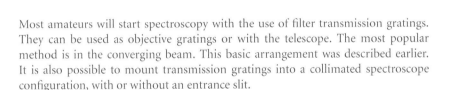

The filter grating can easily be mounted into a suitable filter ring and positioned in front of the camera lens. This will limit the aperture to approximately 25 mm but will still show spectra of the brighter stars. The dispersion/plate scale can be determined from equations presented earlier.

Larger transmission gratings can also be used and mounted in a similar manner to the prisms. See Figs. 14.1 and 14.2.

Converging Beam Arrangement

Earlier we explained how to set up and use the filter grating. Here we will describe in a bit more detail how the spectrum is formed, and later, how to improve the performance.

K.M. Harrison, *Astronomical Spectroscopy for Amateurs*, Patrick Moore's Practical
Astronomy Series, DOI 10.1007/978-1-4419-7239-2_14,
© Springer Science+Business Media, LLC 2011

Figure 14.1. SA100 objective grating mounted on 200 mm telelens.

Best results are obtained with smaller f/ratios, which give smaller star images. For any particular grating, the linear dispersion/plate scale is proportional to the distance from the grating to the film plane.

NOTE: The spreadsheet TransSpec V2 (See Appendix D in this book) is an ideal resource for calculating the various parameters for a filter grating placed in the converging beam. See Fig. 14.3.

Figure 14.2. P-H objective grating mounted in a Cokin filter holder.

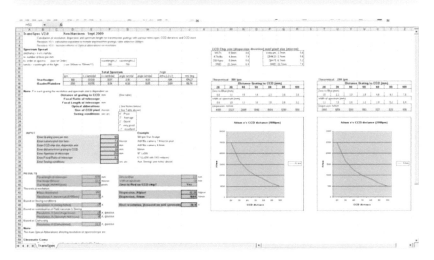

Figure 14.3. TransSpec input screen.

Similar calculations can be done using the following routine:

$$nN\lambda = \sin\alpha + \sin\beta$$

defines the relationship between the angle of the incident beam (α), the angle of the exit (or reflected beam) (β), the spectrum order (n), the grating l/mm (N), and the wavelength of the light (λ).

Using the grating equation with an entrance angle, $\alpha = 0$, we get:

$$Sin\beta = n\,N\,\lambda$$

The spectral deviation of the first order spectrum ($n = 1$), a 100 l/mm grating (N) based on 550 nm (λ) is:

$$Sin\beta = 1 * 100 * 550 * 10^{-4}$$

$$\beta = 3.15°$$

Similarly we can find the deviation angle for the blue and red regions of the spectrum:

Blue, λ_b, 380 nm $\beta_b = 2.18°$ (0.03815 radian)
Red, λ_r, 700 nm $\beta_r = 4.01°$ (0.07017 radian)

This gives a dispersion (D) of:

$$\text{Dispersion (D)} = (\lambda_r - \lambda_b)/(\beta_r - \beta_b)\ \text{Å/degree}$$

$$= (700 - 380)/(4.01 - 2.18)$$

$$= (320)/(1.83)$$

$$= 174.86\ \text{nm/degree}$$

$$= 9{,}992\ \text{nm/Radian (99,920 Å/radian)}$$

(Note: 1° equals 0.0175 radian)

To determine a plate scale we need to select a distance (L mm) between the grating and the CCD.

$$\text{Plate scale, Å/mm} = D/L\ (D\ \text{in Å/radian})$$

Based on a distance of 80 mm we get:

$$\text{Plate scale} = 99{,}920/80$$

$$= 1249\ \text{Å/mm}$$

This plate scale can also be expressed in terms of the camera pixel size, in μm:

$$\text{Plate scale, Å/pixel} = (0.001\ D/L)\ \text{pixel size}$$

where D is in Å/mm.

Assuming a pixel size of 7.4 μm:

$$\text{Plate scale, Å/pixel} = (0.001 {}^{*}1{,}249){}^{*}7.4$$

$$= 9.24 \text{ Å/pixel}$$

For a 100 l/mm grating, used in a 200 mm SCT at f6.3 and a grating distance of 80 mm with a DSLR (7.4 μm pixels), the plate scale is 1,249 Å/mm or 9.24 Å/pixel.

Distances of 30–80 mm are regularly used. If the CCD sensor is large enough, it's possible to record the star's zero image and its spectrum at the same time. This makes subsequent analysis easier.

As the grating is positioned in a converging light beam the focal point will be shifted outwards towards the back of the grating by a third of the thickness (You see a similar effect when using glass filters in filter wheels close to the telescope focus). This can mean a difference in focus position with or without the grating of up to 1 mm. The resolution is restricted by the size of star image, telescope optical aberrations, distance to CCD, and the number of lines illuminated. The size of the star image effectively becomes the entrance slit, and obviously the smaller the image the better the resolution.

$$\text{The linear star image size} = \tan(\text{seeing}) {}^{*}F$$

Where seeing is expressed in arc seconds, F is the effective focal length of the telescope. Typical seeing will be between 1″ (excellent!) and 4″ (pretty poor).

The following optical aberrations affect the resolution when the grating is used in the converging beam:

Chromatic Coma

This aberration is similar to that seen in fast optical systems where "seagull" like distortions are evident towards the edge of the field of view, and affects all wavelengths of light.

The formula for chromatic aberration is:

$$\Delta\lambda = 3\lambda/8f^{2}$$

As the distorted image contains 80% of the light in 50% of the image, we can safely use half of the above result. Note that the coma varies both with wavelength and the inverse square of the focal ratio (f). A faster focal ratio therefore gives smaller star images but more chromatic coma.

In terms of resolution:

$$\text{Linear comatic image size} = 3\,L\,\tan\beta/16\,f^{2}$$

$$\text{Comatic resolution, } R_c = 16\, f^2/3$$

For an f6.3 system, Hα (6563 Å), and distance between grating and CCD of 60 mm, we get:

$$\text{Comatic resolution, } R_c = 212$$

and a coma image of 31 Å.

Using a grism arrangement can significantly reduce the effects of chromatic coma.

Field Curvature

As the image of the spectrum is formed along a cylindrical curve with a nominal radius equal to the distance between the grating and the CCD, the blue and red regions will be focused at different axial distances from the grating. See Fig. 14.4.

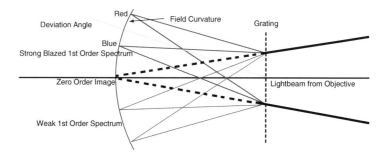

Figure 14.4. Deviation angle and field curvature for a filter grating in a converging beam.

In a setup where the camera/CCD is mounted axially behind the grating, the effects of field curvature can be minimized by focusing the CCD on the spectrum rather than the zero image. For maximum resolution, the particular area (or line feature) should be focused rather than a "nominal" midpoint on the spectrum. The difference between the zero order focus and wavelength λ is:

$$\text{Change in focus, } \Delta L = 3\, L\, (\tan\beta)^2/2$$

where β is the dispersion angle for the target wavelength.

For Hα (6563 Å) (β=3.76°) a 100 l/mm grating, and a grating distance of 80 mm; this gives:

$$\Delta L = 3{}^*80{}^*(\tan(3.76°))^2/2 = 0.52\,\text{mm}$$

The linear out of focus effect at the CCD plane (Δx) is

$$\Delta x = L((1/\cos\beta) - 1)$$

giving $\Delta x = 80 * ((1/\cos 3.76°) -1) = 0.16$ mm.

If the camera/CCD can be tilted or rotated relative to the grating (or the grating itself tilted), then the field curvature can be minimized by setting the angle of the grating to the deviation angle of the spectrum.

Astigmatism

This aberration only affects the width of the spectrum and doesn't impact on the dispersion axis. When considering total flux calculations (i.e., limiting magnitudes and signal to noise) it needs to be considered.

$$\text{Linear astigmatic image} = L(\beta^2 - \alpha^2)/f$$

If the grating is tilted to give $\alpha = -\beta$, then the astigmatism is reduced to zero. As β varies with wavelength, the tilt angle α becomes:

$$\text{Tilt angle}, \alpha = -\,n\,\lambda\,N/2$$

Depending on the system, a maximum of $R = 100$ to $R = 200$ (i.e., a resolution of 60–30 Å) can be achieved.

Aperture Mask

In crowded star fields, the spectrum of the target star may end up lying on an image of a background star (Fig. 14.5). Rotating the grating/camera angle can usually resolve the problem (Fig. 14.6).

Another alternative is to use a small mask with an aperture hole positioned off-center. This allows the target star image to enter the grating but blocks off most of the other field stars, giving a cleaner spectrum.

The mask can be a simple as a 25 mm cardboard disk with a 6 mm hole positioned 3 or 4 mm off center. This is then positioned in front of the grating with small masking tape tabs. A more permanent version can be machined from a piece of aluminum bar (Fig. 14.7). The optimum hole aperture size is:

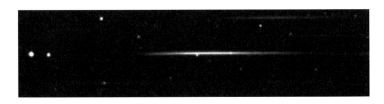

Figure 14.5. SA100 with background star in spectrum. (C. Buil.)

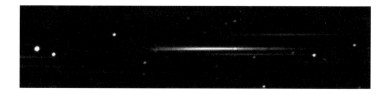

Figure 14.6. SA100 with rotated grating to remove background star. (C. Buil.)

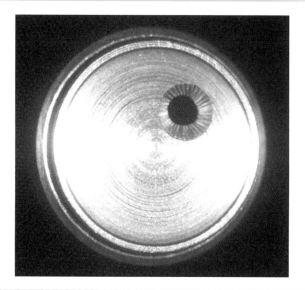

Figure 14.7. Aperture mask. (J. Nott.)

Aperture = (distance between the grating/CCD)/focal ratio of the telescope.

For a grating placed 60 mm in front of the CCD, in an f10 beam, the aperture would be 6 mm.

Adding a Grism

The effects of chromatic coma can be significantly reduced by adding a small prism just in front of the grating to compensate and apply a reverse deviation, bringing the image of the spectrum back onto the optical axis. This configuration is called a "grism" (grating prism), or carpenter prism. The required deviation angle of the prism should be calculated to match the deviation angle of the grating at some nominated wavelength (see Fig. 14.8).

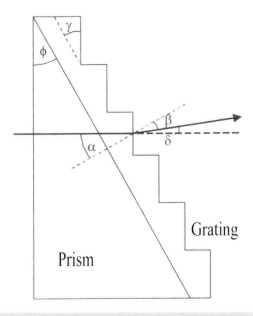

Figure 14.8. Grism optical arrangement. (J. Allington-Smith.)

The basic grating formula is modified by the use of the prism and becomes:

$$nN\lambda = n_o \sin\alpha + \sin\beta$$

where n_o is the index of refraction at the nominated wavelength. As the required deviation $= 0$, $\beta = -\alpha = \psi$

$$nN\lambda_o = (n_o - 1)\sin\psi$$

where ψ is the apex angle of the prism.
    Undeviated wavelength λ_0

$$\lambda_0 = \sin\theta(n_0 - 1)/Nn$$

For the green wavelength, 550 nm

$$550/10^{-7} = \sin\psi(1.5 - 1)/100{*}1$$

$$0.00055 = 0.005\text{Sin}\psi$$

$$\text{Sin}\psi = 0.11$$

$$\psi = 6.3°$$

The required prism deviation angle (θ) can also be expressed as:

$$\theta = n\,N.\lambda_0.n_0/(n_0^2 - 1)$$

For $n_0 = 1.5$, and a first order spectrum ($n = 1$) from a 100 l/mm grating ($N = 100$) this gives:

$$\theta = 120\,\lambda_0\,.\text{radian}$$

So, for the 550 nm wavelength, the prism deviation angle would be:

$$\theta = 120^*550^*10^{-7}.\,\text{Radian}$$

$$= 0.066\,\text{Radian}$$

$$= 3.78°$$

This gives a prism deviation angle, approx. $= 3.8° \times$ l/mm/100

Using this prism will bring the 550 nm wavelength back onto the optical axis. If at the same time the assembly is tilted (see earlier) then both the chromatic and astigmatism aberration are removed. The residual field curvature can be compensated by re-focusing the wavelength under examination.

Suitable 25 mm diameter wedge prisms (Figs. 14.9 and 14.10) are available from most of the optical suppliers (See Appendix A in this book.)

The wedge angle of the prism must be aligned to the direction of the grating grooves to achieve the necessary chromatic corrections.

Unfortunately, adding the prism also introduces some slight non-linearity into the spectrum. For accurate calibration results at least three reference lines must be used.

Adding a Slit

It would be advantageous to be able to add a slit to the filter grating; this would allow you to record spectra from extended objects such as nebulae and galaxies. Other than constructing a full-blown transmission grating spectroscope, you can use the eyepiece as a projection lens. This is achieved by focusing an eyepiece onto a slit at

Figure 14.9. Wedge prism and grating. (Blackwell.)

Figure 14.10. Grism mounted to SA100 grating. (C. Buil.)

a distance of twice the focal length and then projecting the image of the slit through the eyepiece, through the filter grating, and onto the CCD (see Fig. 14.11).

We can use standard adaptors to position a slit even further in front of the eyepiece. For a 25 mm eyepiece, at a 1:1 projection, this distance is 50 mm.

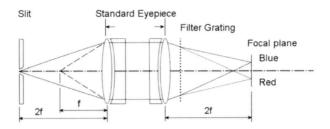

Figure 14.11. The slit mounted at twice the focal length of the eyepiece (f) and the position of the spectrum image.

Using a thin card or sheet metal we can quickly and easily make a slit diaphragm (two pencil sharpener blades) that can be positioned at the front of a filter ring. An adjustable slit can also be used (see below).

By using off-the-shelf adaptors, they can be configured as shown:

• 1.25″ to T thread adaptor, holding the slit assembly.
• Cheap camera adaptor (eyepiece projection) with clamp screw for eyepiece (female T thread at telescope side and male T thread at camera side), holding a 25 mm eyepiece
• T thread to 1.25″ adaptor
• 1.25″ to T thread nosepiece to hold the filter grating.

By positioning the filter grating into the nosepiece of the webcam/CCD and then screwing the slit to into the 1.25″ to the T thread adaptor at the front of the camera adaptor, we can get a projected image of the slit onto the webcam CCD, etc.

Note: The Surplus Shed adjustable slit mechanism (# 1570D) can be fitted into a 15 mm wide T thread spacer. A slot has to be cut to give clearance for the adjusting screw, and the inside diameter needs to be slightly increased (using a small drum sander). Figures 14.12 and 14.13 shows such an assembly fitted to the front of the camera adaptor.

A similar result can be obtained using an Afocal camera adaptor to hold the camera body and grating behind the eyepiece.

Mounting Other Transmission Gratings in a Converging Beam

The commercial transmission gratings mounted in 35 mm frames can be mounted (and used with a grism) in a suitable adaptor. This allows the use of different l/mm gratings.

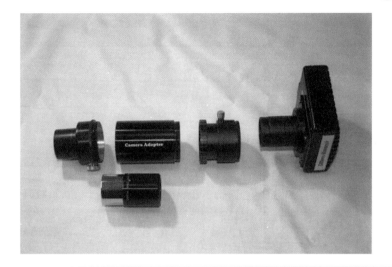

Figure 14.12. Components for adding a transmission grating slit.

Figure 14.13. Grating with slit assembly.

NOTE: The 35 mm framed gratings usually have a smaller grooved grating surface; the mounting adaptor should form an aperture stop to prevent extraneous light from getting through and contaminating the spectrum.

A typical mounting frame is shown (Fig. 14.14) below. This was machined from structural nylon (Delrin) and has a T thread at the front, a pocket for a 25 mm diameter prism (grism), provision for a slide gap to accommodate the grating, and a T thread at the rear. Standard nosepieces and T adaptors can be fitted to suit the telescope and camera.

Figure 14.14. Grating/grism DIY adaptor.

Improving Resolution with a Collimator Lens

Similar to the concepts above, we can add a collimating lens in front of a transmission grating and a camera lens/CCD mounted behind the grating. With the right selection of optics it is possible to fully illuminate the grating and maximize the lines illuminated. Without a slit, the resolution of this arrangement will be limited by the star image size. The collimator lens, grating, and imaging lens should be positioned as close as possible to minimize vignetting. An example of this design is shown in Fig. 14.15.

Figure 14.15. Transmission spectroscope. (I. Baker.)

For first order (n = 1), a 100 l/mm grating (N=100) and green light (λ = 550 nm) the deviation angle is 3.15°. And the theoretical resolution based on a clear grating aperture of 25 mm will be:

$$R = nN = 1^*25^*100 = 2800 \text{ (approx. 2.15Å in the red region)}$$

Ideally the focal ratio of the collimating lens should match that of the telescope and the diameter be large enough to cover the whole grating. Typically a 28 or 30 mm f6 achromatic lens will do the job. The lens needs to be mounted in a suitable tube, or securely on a support board and be able to focus on a star image at the prime focus of the telescope. A grism can be incorporated to bring the spectrum back onto the optical axis and standard camera lens or another achromatic doublet can be used as the imaging lens.

Using a Barlow Lens as a Collimator

This is similar to the arrangement shown in Fig. 13.4 but with the Amici prism replaced by a grating. In this configuration the grating forms a "virtual" objective grating. The aberrations found in the converging beam arrangement are almost removed, and the use of an imaging lens maximizes the dispersion/plate scale and improves the resolution. Adding a grism will reduce the deviation angle. Again a standard camera lens or achromatic doublet can be used as the imaging lens.

You can quickly and easily measure the negative focal length of a Barlow lens by measuring the lens aperture and drawing a circle on a white card double the diameter. For example, if the Barlow lens measures 28 mm, make the circle 56 mm. Project an image of the Sun (which will appear as a large bright disk) onto the card and move it back and forth until the image just fills the circle. Measure the distance to the Barlow lens. This measurement equals the negative focal length of the lens.

Most ×2 Barlows seem to be around −120 mm focal length; this is the distance the lens should be placed inside the prime focus of the telescope to give a parallel collimated beam output.

Watkis Transmission Grating Spectroscope

C. J. Watkis, in 1976, presented a compact design for a transmission grating spectroscope. See Fig. 14.16.

Based on using old binocular objectives, and two (or three, depending on the camera layout) front surface mirrors this design is both compact and functional. Transmission gratings of various size and line density can be used (as can the filter gratings). Provision for rotating the dispersed beam allows the use of high l/mm (600 l/mm) gratings. Depending on the size of the mirrors, it also has the possibility of guiding on the zero image, as shown. The original instrument weighed only 650 gms. Figure 14.17 shows a current example of this design.

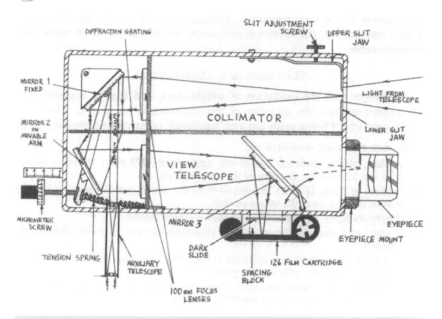

Figure 14.16. Watkis spectrograph – optical layout. (BAA.)

Figure 14.17. Watkis spectroscope. (J. Nott.)

Tragos Design

The Transmission grating optical spectroscope, Tragos was developed by the CAOS team during 2004 for use by students at the Astronomical Observatory of Mallorca. See Fig. 14.18.

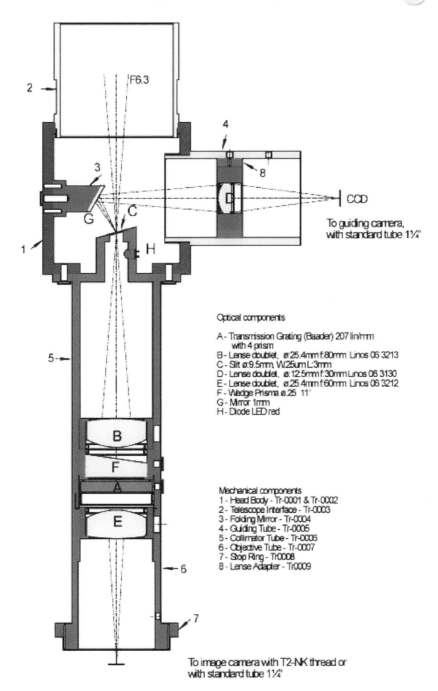

Figure 14.18. Tragos optical layout. (CAOS.)

Optical components

A - Transmission Grating (Baader) 207 lin/mm
      with 4 prism
B - Lense doublet, ø 25.4mm f:80mm Linos 06 3213
C - Slit ø 9.5mm, W.25um L:3mm
D - Lense doublet, ø 12.5mm f:30mm Linos 06 3130
E - Lense doublet, ø 25.4mm f:60mm Linos 06 3212
F - Wedge Prisma ø 25 11°
G - Mirror 1mm
H - Diode LED red

Mechanical components
1 - Head Body - Tr-0001 & Tr-0002
2 - Telescope Interface - Tr-0003
3 - Folding Mirror - Tr-0004
4 - Guiding Tube - Tr-0005
5 - Collimator Tube - Tr-0006
6 - Objective Tube - Tr-0007
7 - Stop Ring - Tr0008
8 - Lense Adapter - Tr0009

To guiding camera,
with standard tube 1¼"

To image camera with T2-NK thread or
with standard tube 1¼"

This elegant design utilizes a filter grating, grism prism, and has additional features such as reflective slit guiding and a rear slit LED illuminator.

LORIS Design

This LOw Resolution Imaging Spectrograph, LORIS shown in Fig. 14.19, is based on an adjustable slit, a 50 mm f1.8 camera lens as a collimator, a 25 mm × 25 mm 300 l/mm transmission grating, and a 35 mm f2 camera lens for imaging.

The imaging lens is inclined to the optical axis at 8.6°, the deviation angle of the 300 l/mm grating. The ratio of the camera lens focal length gives a magnification ratio of ×0.7. For example, a 30 μm slit will project as a 21 μm image on the CCD. This makes it ideal for a 9 μm pixel camera. The adjustable slit allows easy acquisition of the target object, and when used as a "wide slit" (>0.5 mm gap) can signifi-

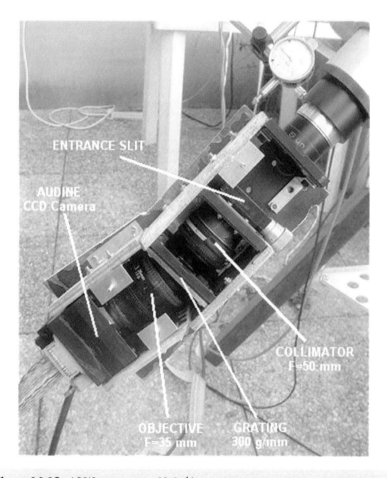

Figure 14.19. LORIS spectroscope. (C. Buil.)

cantly reduce the background illumination. At a SNR >10, a limiting star magnitude of 13.6 was achieved with a 60-min exposure on a 128 mm refractor.

Further Reading

Watkis, C.J. *A Transmission Grating Spectroscope.* JBAA **86** #4, pp. 280–283 (June 1976).

Web Pages

http://www.astrosurf.com/buil/staranalyser/obs.htm
http://astrosurf.com/buil/us/spe1/spectro1.htm
http://users.erols.com/njastro/faas/articles/west01.htm
http://www.britastro.org/iandi/gavin02.htm
http://www.burwitz-astro.de/spectrographs/index.html
http://www.regulusastro.com/regulus/papers/grism/index.html

CHAPTER FIFTEEN

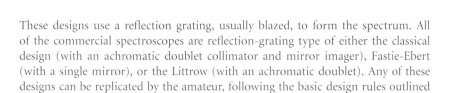

Reflection Grating Spectroscope Designs

These designs use a reflection grating, usually blazed, to form the spectrum. All of the commercial spectroscopes are reflection-grating type of either the classical design (with an achromatic doublet collimator and mirror imager), Fastie-Ebert (with a single mirror), or the Littrow (with an achromatic doublet). Any of these designs can be replicated by the amateur, following the basic design rules outlined earlier.

Based on the practical experience of building a few spectroscopes, having and using an accurate laser pointer during construction is almost mandatory. This device allows you to accurately align the various optics as well as measure slit gaps. Any commercial laser collimating unit should suffice, but it is occasionally found that the devices themselves are not 100% collimated! This needs to be checked and, if need be, corrected before serious use. An alternative is to use a small laser penlight and mount it into a 1.25″ adaptor.

Classic Design

This is probably the easiest spectroscope design for the amateur to construct. It is a combination of an adjustable slit, an achromatic doublet for a collimator, a 30 mm × 30 mm grating, and a camera lens for imaging the spectrum on the CCD.

K.M. Harrison, *Astronomical Spectroscopy for Amateurs*, Patrick Moore's Practical Astronomy Series, DOI 10.1007/978-1-4419-7239-2_15,
© Springer Science+Business Media, LLC 2011

Construction Notes

Weight is always an issue in spectroscope design. A typical spectroscope can weigh (with CCD camera) up to 4 kg. This means counterweights and re-balancing of telescopes and drives. Construction materials can vary from a 2 mm aluminum sheet, Perspex, through to 6 mm plywood and MDF chipboard. As long as the optics can be held rigidly in place with minimal flexing, anything goes. The design discussed here uses 6 mm plywood for rigidity and ease of working. The final overall weight came in at 1.9 kg.

A 60 mm × 60 mm box section arrangement was used, overall sizes to suit the available lenses.

To mount the spectroscope on the telescope a T ring adaptor was screwed onto the front of the spectroscope (This also allows a 2″ T adaptor to be added to suit the scope). This gives a good rigid connection. The slit mechanism (Surplus Shed #1570D) is held in a piece of 18 mm MDF, which allows some radial fine adjustment. A small grub screw locks it in place. The adjustment knob on the adjustable slit doesn't protrude very much, so an extension shaft with a 50 mm diameter disk was added to both seal the area and give fine control of the slit adjuster screw. The exterior disk was subsequently calibrated to give precise settings of the slit gap. A 40 mm binocular lens (collimator) was mounted centrally in a 6 mm divider.

The 30×30 grating is mounted in a holder machined from a length of 20 mm aluminum bar with an 8 mm shaft. This is fitted into an 8 mm precision ball bearing in the base to give good alignment and smooth operation. The actual grating is clamped between a couple of small aluminum angles to allow it to be easily inserted and removed. Interchangeable 600 and 1200 l/mm gratings are used to give different dispersions. See Fig. 15.1.

To get the camera lens (135 mm f2.8) as close to the grating as possible, an old 49 mm filter holder was used to make an adaptor to screw into the front of the lens; this adaptor was epoxy glued and mounted in a 6 mm divider. The gap between the edge of the divider and the main box was sealed with foam to keep dust out. This construction easily allowed the use of other camera lenses and provided a rigid coupling to the spectroscope.

Double check the alignment of the slit and collimator; this must be as accurate as you can make it. A 25 mm hole drilled in the box behind the grating allows the use of a finder telescope to check the focus of the collimator on the slit.

After "dry" fitting all the components to make sure everything went together, the dividers were glued in place. You should have a matte black interior and your color choice of external paint. See Fig. 15.2.

The lid gives access to the slit, grating, and lenses. By setting the grating to the zero position you'll get an image of the slit in the camera; this allows for fine tuning of the slit position and slit gaps and gives confirmation that everything is aligned and working.

Figure 15.1. Typical reflection grating supports.

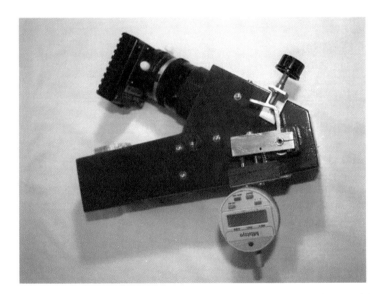

Figure 15.2. Classical spectroscope fitted with a 135 mm f2.8 telelens and Meade DSI II camera.

Grating Adjustments

To improve the accuracy and settings of the grating rotation an external 10 mm ×10 mm × 50 mm aluminum bar was clamped onto the shaft of the grating holder. This bar was rotated (against a small return spring) using a 6 mm threaded rod. On the opposite side of the bar a 3 mm screw was used to set the zero order position of the grating, and a digital dial gauge was mounted to provide a positional readout. This gives an accuracy of 0.01 mm, approximately 7.5 Å/division, with a 600 l/mm grating. See Fig. 15.3.

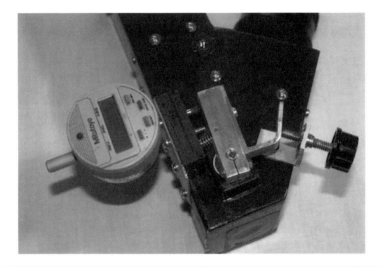

Figure 15.3. Grating adjustment mechanism.

Dust

The biggest problem in real life (other than finding a star!) is dust. ANY dust on the edges of the slit WILL show up in the image. This gives rise to longitude lines in the spectrum –"transversalium" lines. Keeping the slit clean is mandatory. Remember that cleaning the grating is almost impossible, so this should be protected at all times. Likewise any dust on the collimator and camera lenses will affect performance. Try to seal the spectroscope as completely as possible.

Littrow Design

Littrow spectroscopes can be built using an achromatic doublet or a spherical mirror as the collimator. The lens will give some chromatic aberrations, whereas the mirror will introduce coma. The off-axis comatic aberration is doubled by virtue of the second reflection from the mirror.

The Littrow spectroscope design detailed below was built around an achromatic doublet 52 mm f4 lens, using the Surplus Shed adjustable slit and a 30 mm × 30 mm grating. See Fig. 15.4.

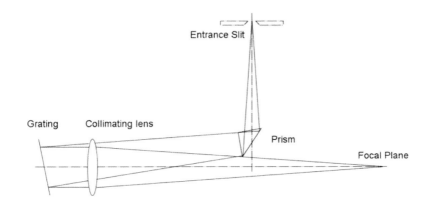

Figure 15.4. Optical layout of the Littrow grating spectroscope.

Construction Notes

The basic "box" is again 60 mm × 60 mm × 6 mm custom wood MDF, with 16 mm sheet used for the various supports. The overall dimensions are 200 mm × 200 mm × 80 mm and weighs in at only 1.2 Kg.

The incoming beam is directed to the collimating lens via a 45° prism (an old binocular prism) positioned just above the optical center line of the lens. The axis of the grating is slightly tilted to cause the return beam to follow the optical axis back to the camera.

A diaphragm aperture of 35 mm × 35 mm was included along with the collimating lens, which fully illuminated the 30 mm × 30 mm grating. This stop effectively turned the instrument into an f6 system, ideal for an LX200 (with the ×0.63 reducer).

Using a 300 l/mm grating, this arrangement gave a very bright spectrum that easily fit the field of most CCD's. The theoretical resolution was about 2.5 Å, more than sufficient to identify the stellar absorption lines for the various stellar classifications.

This is any easy instrument to use and fits well with the smaller telescopes, such as the ED80.

The telescope connection uses a T2 adaptor ring. To ensure minimum distortions, etc., the optical axis from the camera must be kept square to the collimator lens and the grating; the target was to be within 0.1 mm. The ability to easily focus the various cameras (and be able to use various power eyepieces for visual work) was paramount. Positioning the grating required a good shaft support and a method of angular adjustment to align the slit as well as a smooth rotation of the grating with an indicator of position.

The inclusion of a reference light source (neon) was necessary to improve the final calibration of the spectra.

Here, now are some key construction points to keep in mind:

1. Use a robust, lightweight, dust, tight and dimensionally accurate box frame.
2. Have the ability to focus all cameras/eyepieces.
3. Use an improved grating holder and index.
4. Make sure a reference light source is included

The construction is straightforward and can be done using the usual collection of hand tools available to the amateur (Access to a pedestal drill press and a Mitre saw table will help!).

A digital vernier was used to measure each piece of the assembly, to get the accurate positioning of the holes for the mounting ring, slit assembly, collimating lens, grating housing, and the camera/focuser. Measure three times and before cutting, double check again!

An adjustable wood hole drill provides the flexibility of drilling the correct size holes for the various elements (trial holes being cut in scrap until the final dimensions are correct).

The only "special" components, which were machined from aluminum scrap, were the bearing housing for the grating and the grating holder itself.

Following are some additional things to know.

Grating Housing and Adjustment

An 8 mm piece of bright mild steel bar was used as the shaft and an 8 mm precision bearing (around $5) was press fitted into the machined housing. The grating support was milled from a 25 mm diameter piece of aluminum bar with the land for the grating off-set by 6 mm (the thickness of the reflection grating) to bring the front face of the grating onto the center line. Later a 6 mm × 10 mm aluminum tab was epoxy glued to the back of the holder to allow the use of a micrometer head (1″ micrometer head, $4 on eBay) to adjust the position of the grating. See Fig. 15.1.

To minimize errors in adjustment a 6 mm ball bearing was epoxy glued on the nose of the micrometer shaft to act as a bearing point on the tab.

The grating has to be tilted (in the vertical axis) by about 4° to get the primary beam from the slit/prism to return along the optical axis back through the collimator to the camera. This is achieved and adjusted using a push/pull screw arrangement on the bearing housing. The final adjustments were done using the laser collimator.

It was also found that when the grating holder was rotated between the zero image position and the spectrum, that the image on the CCD would appear to move downwards in the frame. For example, if the zero order was located on the center of the CCD (Fig. 15.5, grating position A), then by the time the red portion of the spectrum was brought into view (Fig. 15.5, grating position B) it would be very close to the bottom of the frame. This is caused by the grating axis being inclined to the vertical and the alignment of the grating "tilting" as it is rotated. The solution to this "cylindrical distortion" is to rotate the grating *within* the holder by approximately the same angle as the tilt of the holder.

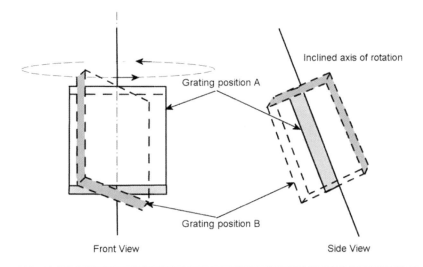

Figure 15.5. Change in grating alignment due to the inclined axis of rotation.

Camera/Focuser

After much thought and many trials with PVC pipe fittings, etc., a Baader T2 rotational adaptor (#BA2456320) and a T2 to 1.25″ micro-helical (4 mm travel) focuser (#BA2458125) were selected as the best solution.

Not cheap but very, very effective (38 € each), these adaptors allow the spectroscope to be focused for the DSLR, which needs 55 mm back focus (worst-case scenario) and then by inserting the micro-focus unit and T2 spacers, the other camera/eyepieces can be mounted and brought to focus independently. The T2 rotational element allows the camera to be orientated and aligned to the spectrum without changing the focal position.

Reference Light Source

Both a neon indicator bulb and an Osram Fluoro energy saving bulb (Duluxs-tar) are used to provide reference spectral lines. Previously they were external sources not incorporated into the spectroscope. To achieve maximum accuracy the reference source must be imaged using exactly the same grating position and camera, so obviously it is far more convenient to have the reference light inside the spectroscope and take an image before and after a spectral exposure.

The neon is the most "recognized" standard, and all the emission lines are fully documented. The small neon bulb is located in front of the slit, and a 9 mm prism is positioned to throw the light on the bottom section of the slit. When the bulb is switched on, it provides a "reference" spectrum in the camera. See Fig. 15.6.

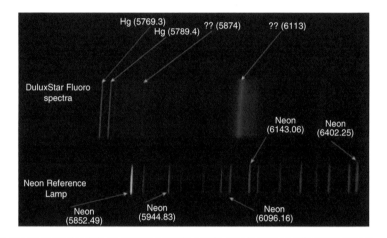

Figure 15.6. Built in neon reference lamp and comparison fluoro spectrum.

The back focus requirements of the Canon DSLR camera automatically determines the distance to the slit.

To provide access to the slit and reference light, the side panel is removable; the grating and collimating lens can be accessed through the top panel of the collimator section.

After sub-assembly and painting, the laser collimator was once again used to confirm everything was correctly aligned; the collimating lens support block was then glued in position. The slit was positioned and the prism angle double checked. The grating holder was then inserted and aligned to the center of the camera adaptor with the laser. Using a 10 mm eyepiece positioned 55 mm behind the T2 camera adaptor, the slit was adjusted to perfect focus, then the slit mounting block glued in position. The final position was double checked by

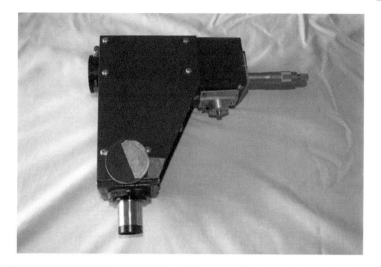

Figure 15.7. Final Littrow spectroscope.

setting up the Canon DSLR and taking some exposures to confirm alignment and focus.

The slit was aligned to the grating by eye and locked into position. The "zero order" position was established on the micrometer and verified by moving the grating through its full travel back to the zero position (see Fig. 15.7).

Slit Back Light Arrangement

After the initial trials of the spectroscope it was quickly found that positioning the star image central to the slit was difficult. Sometimes it would be positioned at one extreme or the other. To make this job easier a red LED illuminator (from a crosswire eyepiece) was added to the body of the spectroscope, just above the prism. See Fig. 15.8. This shines onto a polished aluminum angle that directs the beam onto the back of the slit. When the slit is illuminated it is very easy to see the position of the star image on the slit. See Fig. 15.9.

WPO Design

The Worcester Park Observatory (WPO) spectroscope, designed by Maurice Gavin in 2000, is a very elegant, compact Littrow design based a telephoto lens as the collimator. See Figs. 15.10 and 15.11.

In the original design, the return beam to the CCD is shown positioned below the optical axis, and due to the constraints of the available adaptors this is usually

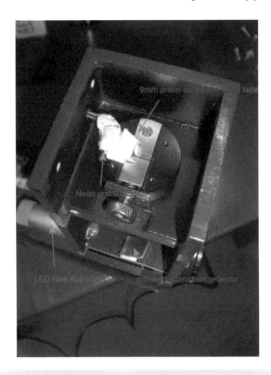

Figure 15.8. Arrangement of the back light and neon reference light.

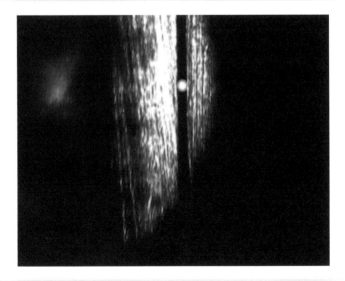

Figure 15.9. Back illuminated slit and star.

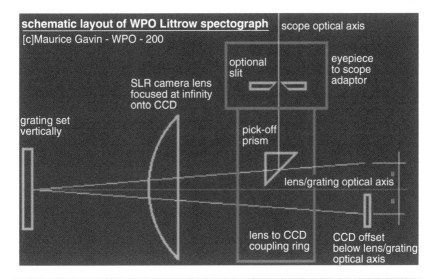

Figure 15.10. WPO optical layout. (M. Gavin.)

changed to bring the CCD onto the optical axis, which can be achieved by slightly tilting the axis of grating. Using standard camera lenses at the infinity setting (to give a collimated beam for the grating) restricts the available back focus to 55 mm, so some careful thought has to be given to the placement of the pick-off prism and the size of the adaptor body. The f/ratio of the camera lens is usually very low (a 135 mm telelens needs to be set to at least f4) to fully illuminate the grating.

A variation on the WPO design is shown in Figs. 15.12 and 15.13.

Norbert Reinecke, in his version of a compact Littrow (Fig. 15.14), used standard adaptors and reversed the photographic lens, i.e., mounted the lens with the "rear" element facing the grating. This gives an increase in the available back focus. A needle is also used to reflect a reference lamp into the light path.

Peter Kalajian has also built a compact Littrow spectroscope using a standard camera lens. The grating support frame and pick-off prism housing were designed and verified using CAD before manufacture and assembly. The CAD designs are illustrated in Figs. 15.15 and 15.16.

Ebert-Fastie Design

This design uses a single mirror, usually spherical, as both the collimator and the imaging lens. The coma from the first reflection is canceled by the second. The minimum mirror diameter has to be chosen to suit the size of the grating and will be approx 3.6 times the grating width. The grating rotation axis should be positioned on the optical axis of the mirror, and the light from the slit can be introduced via a

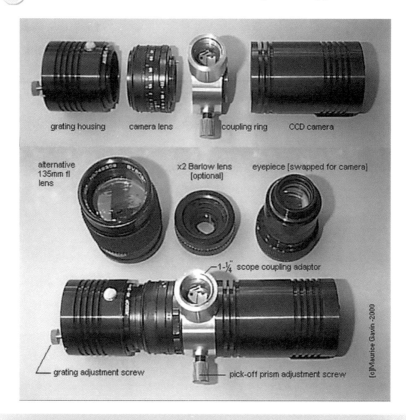

Figure 15.11. WPO components. (M. Gavin.)

Figure 15.12. Compact WPO components.

Figure 15.13. Compact WPO spectrograph.

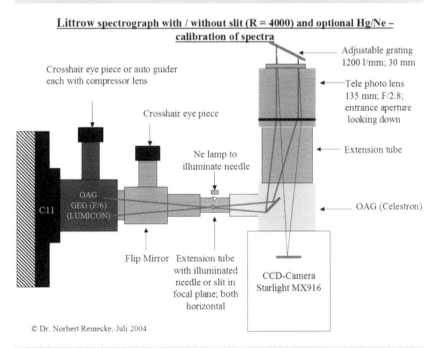

Figure 15.14. Reinecke's compact Littrow spectrograph. (N. Reinecke.)

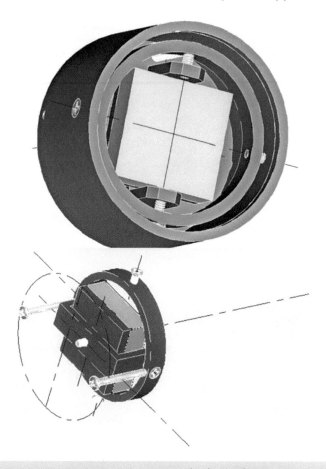

Figure 15.15. Compact Littrow grating mounting details. (P. Kalajian.)

small mirror. A similar mirror can be used to direct the diffracted beam to the CCD. This arrangement allows the spectroscope to be more compact and rigid.

The f/ratio of the mirror must obviously suit the telescope. A 125 mm f6 (depending on the telescope) mirror would be right for a 30 mm grating. This gives a very long (750 mm) focal length and is therefore difficult to mount securely on the telescope.

A fiber optic feeding the slit, with the spectroscope mounted remotely, would probably be a better solution. The focal ratio of the mirror could be reduced to suit the NA of the fiber optic (see below).

Focusing the camera/CCD can be achieved by either a conventional focuser at the camera attachment, or by moving the mirror along its axis (This is the method used in the SBIG SGS spectrograph).

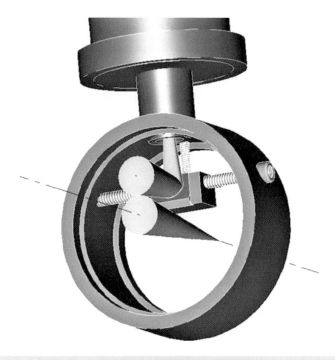

Figure 15.16. Compact Littrow design. (P. Kalajian.)

Czerny-Turner Design

Similar to the Ebert-Fastie design but uses separate mirrors for the collimator and the imaging lens. This allows the use of different mirror focal lengths to give a magnification factor and freedom of positioning of the imaging mirror to improve the anamorphic factor. Coma is not fully corrected, as is the case in the Ebert-Fastie design, but by tilting the imaging mirror it can be significantly reduced.

The individual mirrors must be large enough to provide a beam diameter covering the grating. See Fig. 15.20.

Echelle Spectroscopes

Echelle spectroscopes use a very coarse high efficiency blazed grating, unlike the traditional spectroscopes that use high l/mm reflection gratings, and works in very high spectral orders. The required incident angle is very extreme and close to the blaze angle (usually 50–70°)

Echelle spectroscopes can be used in spectral orders well above n = 40, to give high dispersions over a short wavelength range. When used in these high spectral orders the free spectral range is very low (<100 Å), and the overlapping spectra are separated using a second grating (or prism), called the cross disperser, positioned at right angles to the echelle. By using these high spectral orders very high resolution (R>50,000) can be achieved. Standard lamps can be used to calibrate the spectrum, but the matrix of short spectral sections requires sophisticated software to do the final analysis.

$$\sin\beta = nN\lambda$$

It follows that in higher orders the angular separation between two wavelengths becomes greater. Consider two lines, one at 600 nm and the other at 605 nm, incident on an echelle grating with 79 l/mm. From the equation above, at n = 1, the angular separation is 0.03° but at n = 40 the angular separation increases to 2.1°. The disadvantage is the reduced free spectral range, which decreased from 630 nm (630 nm/1) to only 15.6 nm (630 nm/40).

Although challenging and difficult to accurately construct, the fixed echelle grating and cross disperser is attracting attention within the ranks of the advanced amateur, and prototypes have already been constructed and tested in Europe. The Forum der VdS-Fachgruppe Spektroskopie contains useful information for constructors (In German).

Fiber Optics

Many advanced amateurs are now using a fiber optic link between the telescope and the entrance slit (sometimes the fiber replaces the slit). This arrangement allows the spectroscope to be located in the observatory remote from the telescope, giving more freedom to the weight and design of the spectroscope as well as obviating the usual problems of rigid attachment and balancing the telescope/spectroscope assembly.

A fiber optic link is designed to allow, by total internal reflection, the transfer of a signal, in our case light along its length. This is achieved by coating a core optical fiber with a sleeve material of a different refractive index. Typical core diameters are 9 μm and 50 μm, with final outside diameters of 125 μm. This glass tube is then protected by being encased in a PVC outer sleeve usually 3 mm diameter, which also allows various end fittings to be attached.

Fiber links are now available (ex audio connectors) that provide polished square ends suitable for the amateur spectroscope builder. These come in various lengths from 1 to 5 m.

Aperture Ratio, F/Ratio, and Numerical Aperture (NA)

More commonly used to describe microscope objectives than telescopes, the numerical aperture (NA) is a key measure of the light-gathering power of an objective or a lens (NA is widely used for f/ratios below f2):

$$NA = nSin\theta$$

where n is the refractive index of the material and θ the half angle of the beam from the lens.

This can be translated to our more common f/ratio:

$$f/ratio = 1/2NA$$

Therefore a f2 lens would have a NA = 0.25 and a f5 lens NA = 0.1 The minimum f/ratio transmitted by a fiber optic is:

$$f\,min = \frac{\sqrt{1 - NA}}{2NA}$$

Fiber Optic Applications

Using a fiber optic as an entrance slit on a telescope gives an entry beam equal to the telescope's f/ratio. Depending on the design of the fiber, some of this light may be lost due to the lack of reflection between the core and the sleeve. Likewise the resulting exit beam will always be more spread out. For example, if an f8 beam is focused onto the end of the fiber, only 55% of the energy will be contained at the output of the fiber at the same f/ratio. However if the input beam is f3, 85% of the light will be in an exit beam of the same f/ratio. This f/ratio degradation gets worst when the fibers are bent or twisted. This needs to be considered in the design of the spectroscope collimator. The overall transmission efficiency can be up to 90% across the visible spectrum.

Using an ×0.33 reducer on a SCT would achieve the best entrance f/ratio; alternatively a transfer lens system can be used to get the faster f/ratio. A similar lens will be required at the exit to bring the beam to the collimator f/ratio.

A fiber optic link can be used in conjunction with the spectroscope entrance slit. The loss of light (if the input beam is corrected to the collimator f/ratio) will be similar to that outlined earlier.

The fiber optic can be simply mounted into the telescope focuser using a 1.25″ plug. Guiding can be done using a beamsplitter or the fiber can be stripped of its

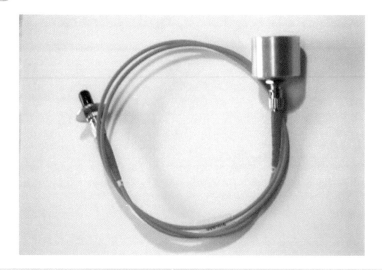

Figure 15.17. Standard fiber optic cable fitted to 1.25″ nosepiece.

protective covering and located in a holder with a reflective surface at 45°; the guide camera then focuses on the star image sitting on the end of the fiber. See Fig. 15.17.

Multiple fibers can be used, aligned in a circle (6 plus one central) or in a linear array to give larger apertures. One option (used by Tom Kaye on this High Precision Radial Velocity Spectroscope) is to have a circular array on the telescope end and map these to a linear array at the exit.

Other Amateur Designs

As with many other branches of amateur astronomy, there's always room for modification and adaptation in the design and construction of spectroscopes. Some alternative designs that may give some food for thought are described below.

Mete

Fulvio Mete has developed a "cylindrical lens amateur universal spectroscope" (CLAUS) where a cylindrical lens is used to present a virtual slit and act as a collimator. See Fig. 15.18.

The initial beamsplitter allows guiding, and the reducer lens (standard ×0.33 reducer) provides a small star image to the focus of a 25 mm diameter, 50 mm focal length cylindrical lens. The collimated beam then goes to the 25 mm × 25 mm, 1200 l/mm grating and the diffracted beam to a 50 mm f2 camera lens for imaging.

The results are impressive, and a sample series of spectra is shown in Fig. 15.19.

C.L.A.U.S. Project
Cylinder lens amateur universal spectroscope
by Fulvio Mete

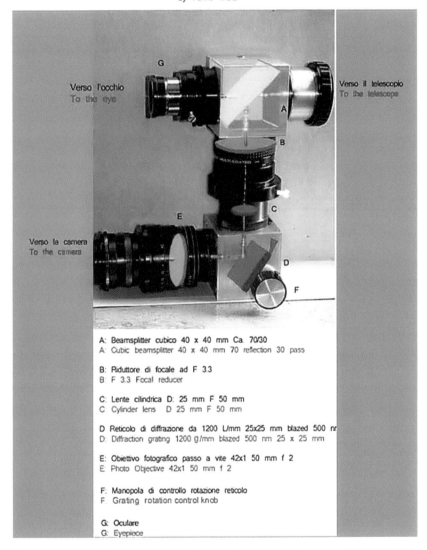

A: Beamsplitter cubico 40 x 40 mm Ca. 70/30
A: Cubic beamsplitter 40 x 40 mm 70 reflection 30 pass

B: Riduttore di focale ad F 3.3
B: F 3.3 Focal reducer

C: Lente cilindrica D: 25 mm F 50 mm
C: Cylinder lens   D 25 mm F 50 mm

D Reticolo di diffrazione da 1200 L/mm 25x25 mm blazed 500 n
D: Diffraction grating 1200 g/mm blazed 500 nm 25 x 25 mm

E: Obiettivo fotografico passo a vite 42x1 50 mm f 2
E: Photo Objective 42x1 50 mm f 2

F: Manopola di controllo rotazione reticolo
F: Grating rotation control knob

G: Oculare
G: Eyepiece

Figure 15.18. CLAUS spectrograph – optical layout. (F. Mete.)

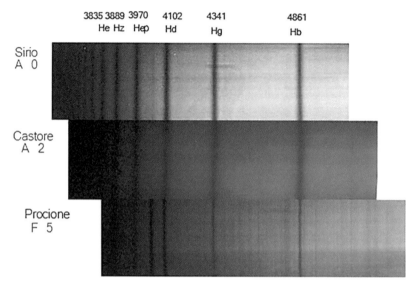

Fulvio Mete 11.12.2004 - CLAUS Spectroscope

Figure 15.19. CLAUS spectra showing the hydrogen absorption *lines*. (F. Mete.)

Kaye

To meet the needs of his exoplanet searches, Tom Kaye designed and built a long focus, high resolution, temperature stabilized spectroscope, fed by a fiber optic link from his 16″ telescope. See Figs. 15.20 and 15.21. This Czerny-Turner spectroscope

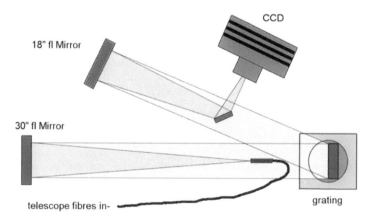

Figure 15.20. High precision spectrograph – optical layout. (Adapted from T. Kaye.)

Figure 15.21. High-precision spectrograph – general arrangement. (T. Kaye.)

utilizes a 150 mm f6 mirror as the collimator, a 100 mm × 100 mm 1800 l/mm grating, and a 150 mm f3 mirror for imaging. All this is mounted on a substantial granite base plate and sits in an insulated thermally controlled enclosure. This instrument is capable of measuring radial velocities of 200 m/s!

Figure 15.22. Compact Research Spectrograph. (N. Glumac.)

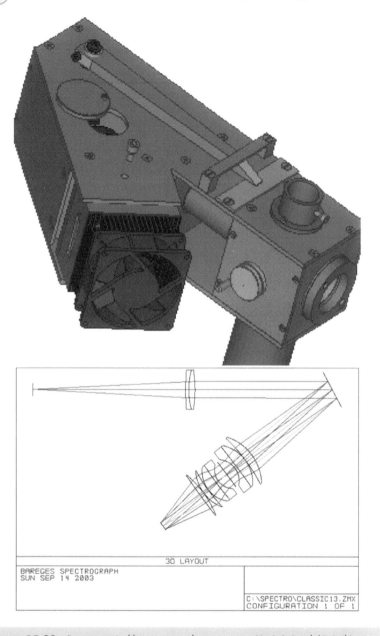

Figure 15.23. Bareges optical layout, general arrangement. (A. & S. Rondi/C. Buil.)

Glumac

Nick Glumac has build a few spectroscopes, one of which, the Compact Research Spectrograph (CRS) is of interest.

This is basically a compact classical design spectroscope with a fixed 50 μm slit, using a 15 mm f4.5 achromatic lens as the collimator, a 25 mm × 25 mm, 1,800 l/mm grating, and a 25 mm f1 CCD camera lens for imaging. See Fig. 15.22.

This instrument gives a FWHM line width of 25 μm and a resolution of 5.6 Å

Bareges CCD Spectrograph (Rondi/Buil)

In 2003 A. & S. Rondi with Christian Buil laid down the design of their "Bareges" classical designed spectroscope. This is based on an entrance 75/25 beamsplitter, an adjustable slit, 30 mm f5 achromatic doublet as the collimator, 30 mm × 30 mm, 600 l/mm grating, and a standard 50 mm f1.4 camera lens for imaging. See Fig. 15.23.

Further Reading

Martinez, P. (Ed.). *The Observers' Guide to Astronomy,* Vol. 2, pp. 689–696, Cambridge University Press (1994).

Tonkin, S.F. (Ed). *Practical Astronomical Spectroscopy.* Springer (2002).

Thorne, A. et al. *Spectrophysics Principles and Applications,* p. 257. Springer (1999).

Web Pages

http://www.astro-reinecke.org/html/spectroscopy.html

http://mais-ccd-spectroscopy.com/

http://www.starlink.rl.ac.uk/star/docs/sg9.htx/sg9.html#stardoccontents

http://www.astroman.fsnet.co.uk/newspec.htm

http://www.spectrashift.com/spectrographs_high_reso.shtml

http://www.lightfrominfinity.org/gli_strumenti.htm

http://www.telescope-service.com/baader/adaptors/adaptors.html#Large

http://www.surplusshed.com/pages/item/m1570d.html

http://www.surplusshed.com/pages/item/l2047d.html

http://www.etsu.edu/PHYSICS/aas190/aas190.htm

http://articles.adsabs.harvard.edu//full/1979PASP...91..149S/0000149.000.html

http://www.stsci.edu/hst/stis/documents/isrs/199823.pdf

http://www.astrosurf.com/rondi/spectro/index.htm

http://www.astrosurf.com/aras/bareges/bareges_en.htm

http://mechse.illinois.edu/research/glumac/astro/astro.html

http://www.chara.gsu.edu/CHARA/Reports/appendixy.pdf

http://www.freepatentsonline.com/6628383.html

http://www.itpa.lt/~laroma/nordforsk2008/Lectures/EchelleDataReduction.pdf

CHAPTER SIXTEEN

Guiding, OAG, Beamsplitters, and Flip Mirrors

You'll quickly find it's a challenge to get a star focused on the spectroscope slit and hold it there during a series of 5-min exposures!

The Shelyak LhiresIII and others add a small "telescope" to view the image of the star sitting across the slit gap. Another option is to introduce a beamsplitter in front of the slit to allow setting and guiding. You could use a flip mirror to set up the spectroscope on the star, but you lose the ability to guide.

Off-Axis Guiders

To prevent any additional loss of light, i.e., using beamsplitters, etc., many astronomers use an off-axis guider (OAG) to locate a suitable guide star close to the target star. An OAG design such as the Celestron Radial Guider or the Orion Deluxe OAG give the opportunity of setting the target star on the entrance slit and then adjusting the position of the pick-off prism to locate a guide star.

Beamsplitter Spectroscope Applications

One of the problems using a slit spectroscope is centering and guiding on a star image. Most professional observers use a reflecting slit and secondary optics to present a star image to the guider. Unfortunately these slits are expensive and usually

K.M. Harrison, *Astronomical Spectroscopy for Amateurs*, Patrick Moore's Practical Astronomy Series, DOI 10.1007/978-1-4419-7239-2_16,
© Springer Science+Business Media, LLC 2011

have a fixed slit width (i.e., 30 μm). For amateur spectroscopes, which use an adjustable slit (0–3 mm) (Surplus Shed # 1570D), this method is not suitable. By modifying a flip mirror to hold a suitable beamsplitter, an image of the star can be seen in the guider and at the same time focused on the spectroscope slit.

Beamsplitters come in various forms:

- A beamsplitter cube; These give a 50/50 split of the incoming light, which will reduce the star's brightness on the spectroscope slit by about one magnitude. See Fig. 16.1.

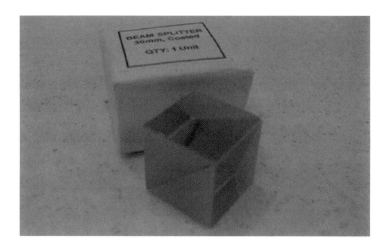

Figure 16.1. Beamsplitter cube.

- A beamsplitter plate. Beamsplitter plates are thin (1 mm) and give a 70/30 split; this reduces the light loss to around 0.5 magnitude. This is a preferred solution (See Fig. 16.2). The beamsplitter plate, as supplied, is 50 mm × 50 mm but can easily be cut to size with a normal glass cutter.

Modifying a Vixen Flip Mirror to a Beamsplitter

Vixen manufactures a nice robust flip mirror, Model # 2680. See Fig. 16.3.

This unit has a built-in 2″ nosepiece with provision for mounting filters; unfortunately it's a 49 mm thread, so a 48 or 49 mm reducer is required to use the standard 2″ astronomical filters.

Figure 16.2. Beamsplitter plate.

Figure 16.3. Vixen flip mirror.

The flip mirror is a 30 × 35 mm front surface mirror mounted on a plate that is moved via a small pin and groove mechanism from a knob on the outside of the body. This gives smooth and positive movement of the mirror. The body is die-cast and approximately 55 mm × 55 mm × 55 mm.

The outlet ports on the body, for both the imaging and guider, have female T threads, and male to male T thread adaptors are supplied as well as the T thread to 1.25″ adaptors. This allows the cameras to be mounted (using the T threads) very close to the body and reduces the back focus requirements.

The following notes explain how to modify a flip mirror and add a beamsplitter. DO THIS AT YOUR OWN RISK. Read through the instructions before starting and make sure you understand the various steps involved. If you are careful, however, it should be easy to reverse the modification and return the flip mirror to its original state.

Before you start to disassemble the Vixen flip mirror, you'll need the following items:

- 2 mm Allen key
- 0.9 mm Allen key
- a Philip's head screwdriver
- an Exacto knife (or similar)
- acetone (or similar)

1. Remove the 1.25″ adaptors from the imaging and guide ports – just unscrew them by hand.
2. Remove the 2″ nosepiece by unscrewing the four Philip's head screws.
3. Remove the adjusting knob; there's a small 2 mm grub screw inside the body that holds it in place. NOTE: The pin mechanism will now drop out; get ready to catch it without touching the mirror!
4. Remove the flip mirror plate. There are two 2 mm grub screws on the outside of the body that form the "axis" for the plate; gently unscrew them and remove the mirror plate. NOTE: Be careful not to touch the front surface mirror!
5. Remove the front surface mirror from the plate. There are four small dabs of glue at the edges of the mirror; these can be removed by carefully using a sharp blade (Exacto knife). The mirror is held in place with a piece of double-sided sticky tape; this can be softened and removed by soaking for a few minutes in Acetone. To assist in lifting the mirror off the tape, you may find you have to use the small (and they are small!) grub screws at the rear of the plate. If you look at the back of the mirror plate, there seem to be four small holes; these are actually 0.9 mm grub screws that can be used to give fine "adjustment" of the mirror position. Gently screwing in one of the grub screws will lift the edge of the mirror and allow it to be removed from the plate. NOTE: The acetone has no detrimental effect on the mirror. Once removed, wrap in tissue paper and stored (You never know when you might need it for another project!).

That's it for disassembly!

To modify the mirror support plate for the Beamsplitter, you'll need:

- a 12 mm drill bit (minimum)
- a glass cutter
- double-sided sticky tape

The mirror support plate needs a hole drilled in the center to allow the secondary light beam to get through.

You have a choice; you can either incline the plate on a jig/support and drill the hole at 45°, which will give you a circular aperture when inclined, or you can take the easy way and just drill the plate flat!

The "easy way" was taken here and the 12 mm hole drilled through the center of the plate. Even when inclined at 45°, the elliptical aperture is still about 7 mm × 12 mm, which was more than enough clearance for the guide camera light beam.

Fitting and Adjusting the Beamsplitter Plate

Cut a rectangular section from the larger piece of beamsplitter, 20 × 34 mm. A normal wheel glass cutter does the job in seconds!

The beamsplitter plate can then be mounted on the front of the support plate (Remember to unscrew the little grub screws, if you used them to lift the mirror!). Put two strips of the double-sided tape, about 6 mm wide, across the top and bottom surface of the beamsplitter. Gently place the beamsplitter in the small recessed area, central to the support plate and hole. The tape should hold it firmly in place.

Re-assemble the mirror plate and the adjustment mechanism. See Fig. 16.4.

When the mirror plate is rotated down, you now have a 70/30 beamsplitter in the optical train. Set it up on your telescope and put the low power eyepiece in both the guide and imaging adaptor. Focus using the guide port (rear port) and center the star.

Figure 16.4. Beamsplitter in place.

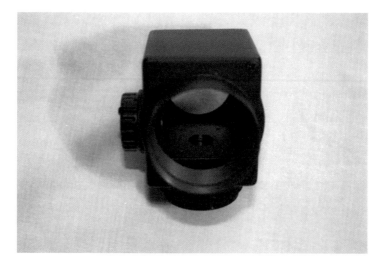

Figure 16.5. Rear view, showing the four small adjusting screws and the hole.

Now look in the imaging/spectroscope port and see if the star image is still central. If it is not, the small grub screws in the mirror support can be used to *slightly* tilt and rotate the beamsplitter plate to bring the star back to the center. See Fig. 16.5.

Using the Beamsplitter

The back focus distances for the guide and spectroscope depend on the equipment you have.

A QHY5 guide camera was used and mounted in a 1.25″ nosepiece. To get correct focus the Vixen adaptor was removed and replaced with an adjustable Baader T thread to 1.25″ adaptor (Baader # BA2458010). The spectroscope has a female T thread, so this was just screwed directly to the Vixen body. See Fig. 16.6.

To set up for taking a spectra, centralize and focus the star in the guide camera/eyepiece. The star should then be visible on the spectroscope slit; this can be verified by imaging the zero order image through the spectroscope.

Dove Prism as Beamsplitter

G. D. Roth, in his *Handbook for Planet Observers*, suggests using a Dove prism as a beamsplitter. He claims that 75% of the light is reflected from the front surface and 25% (20% after losses) goes through to the guide eyepiece. See Fig. 16.7. This design could easily be adapted to a spectroscope.

Figure 16.6. The beamsplitter and QHY5 guide camera in position on a Littrow spectroscope.

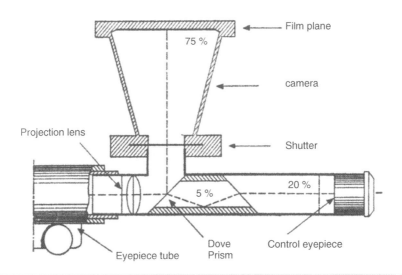

Figure 16.7. Dove prism as a beamsplitter. (G. Roth.)

Transfer Systems for Guide Cameras

A single lens can be used to view the star image on the slit. The simple lens equation can be used to calculate the separation distances. As the lens moves to achieve focus the effective magnification will change. As an alternative two short focal-length achromatic doublets can be used to make a transfer lens system. See Fig. 16.8. This arrangement allows the guide camera to be focused on the star image sitting on the reflective slit.

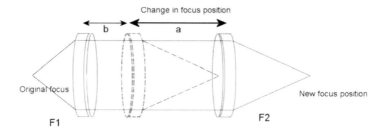

Figure 16.8. Optical arrangement of transfer lens.

The total distance between the slit and the guide CCD is

$$D = F_1 + (a + b) + F_2$$

where F_1/F_2 are the focal lengths of the lenses and $(a + b)$ the distance between them. The focus point can be moved by changing the separation $(a + b)$ between the lenses. This configuration has the added benefit of allowing the distance to the guide camera to be varied without changing the magnification.

Split Mirror Guider

By replacing the beamsplitter mirror with two small front surface mirrors positioned parallel with a 4–6 mm gap, an effective reflection guide system can be made. The star image is reflected to the guide camera without any additional transfer optics.

The actual mirror gap will depend on the focal ratio of the telescope and the distance from the split mirror to the entrance slit of the spectroscope. See Fig. 16.9.

The guide star image is slightly more homogenous than that seen in traditional reflection slit applications. See Figs. 16.10 and 16.11. This "reflective" slit guiding accommodates spectroscopes that use a standard non-reflective adjustable slit.

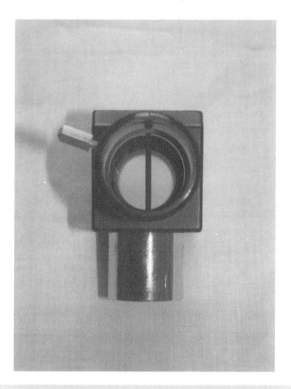

Figure 16.9. Split mirror guider.

Figure 16.10. Image of out of focus star from the split mirror guider.

Figure 16.11. Image of star close to focus from the split mirror guider.

Further Reading

Dragesco, J. *High Resolution Astrophotography.* Cambridge University Press (1995).
Roth, G. D. *Handbook for Planet Observers.* Faber and Faber (1966).

Web Pages

http://www.opticsplanet.net/vixen-flip-mirror.html
http://www.surplusshed.com/pages/item/pm1063.html

Appendix A

Suppliers of Spectroscopes and Accessories

Star Analyser Grating

AVA ASTRO CORP, DBA – Adirondack Video Astronomy, 72 Harrison Avenue Hudson Falls, NY 12839 USA
http://www.astrovid.com/prod_details.php?pid=3490

Paton Hawksley Education Ltd, Rockhill Laboratories, Wellsway, Keynsham, Bristol, BS31 1PG U.K.
http://www.patonhawksley.co.uk/staranalyser.html

Rainbow Optics Star Spectroscope

RAINBOW OPTICS 2116 Regent Way Castro Valley, CA 94546
http://www.starspectroscope.com/

SGS and DSS-7 Spectroscopes

Santa Barbara Instrument Group
147-A Castilian Drive
Santa Barbara, CA 93117

K.M. Harrison, *Astronomical Spectroscopy for Amateurs*, Patrick Moore's Practical
Astronomy Series, DOI 10.1007/978-1-4419-7239-2,
© Springer Science+Business Media, LLC 2011

http://www.sbig.com/sbwhtmls/spectrometer2.htm
http://www.sbig.com/sbwhtmls/online.htm

Shelyak Lhires III

Shelyak Instruments
Les Roussets
38420 Revel
France
http://www.shelyak.com/

Baader Dados Spectrograph

Baader Planetarium
Zur Strenwarte
D-82291 Mammendorf, Germany
http://www.baader-planetarium.com/

Rigel Systems

Rigel Systems
26850 Basswood Avenue
Rancho Palos Verdes
CA 90275 USA
http://www.rigelsys.com/

QMax Spectrograph

Questar Corp.
6204 Ingham Road
New Hope, PA 18938 - USA
http://www.questarcorporation.com/spectro1.htm

Gratings

Optometrics, 8 Nemco Way, Stony Brook Industrial Park, Ayer, MA 01432 USA
http://www.optometrics.com/
Paton Hawksley (see above)
ThorLabs
Newton, New Jersey, 435 Route 206 NorthNewton, NJ 07860 USA
http://www.thorlabs.de/

Newport Corp.-(Richardson Gratings)
705 St. Paul St,
Rochester, NY 14605
USA
http://www.newport.com/products/overview.aspx?sec=124&lang=1033

Prisms

Optometrics (see above)

Edmund Optics, Inc
101 East Gloucester Pike,
Barrington, NJ 08007-1380
USA
http://www.edmundoptics.com/

Surplus Shed, 1050 Maidencreek Road, Fleetwood, PA 19522
USA
http://www.surplusshed.com/

Entrance Slits

Surplus Shed (See above)
Shelyak Instruments (see above)

Lennox Laser
12530 Manor Road
Glen Arm, MD 21057
USA
http://www.lenoxlaser.com/

Lenses and Mirrors

Surplus Shed (see above)
Edmund Optics (see above)
Thorlabs (see above)

Electroluminescent Panels

http://www.earlsmann.co.uk/ELPanels.htm

Appendix B

Useful Spectroscopy Forums and Other Websites

http://tech.groups.yahoo.com/group/astronomical_spectroscopy/?yguid=322612425
http://tech.groups.yahoo.com/group/amateur_spectroscopy/?yguid=322612425
http://tech.groups.yahoo.com/group/Iris_software/?yguid=322612425
http://tech.groups.yahoo.com/group/spectro-l/?yguid=322612425
http://tech.groups.yahoo.com/group/staranalyser/?yguid=322612425
http://spektroskopie.fg-vds.de/
http://www.astrospectroscopy.de/
http://www.threehillsobservatory.co.uk/astro/spectroscopy.htm
http://astrosurf.com/buil/us/book2.htm

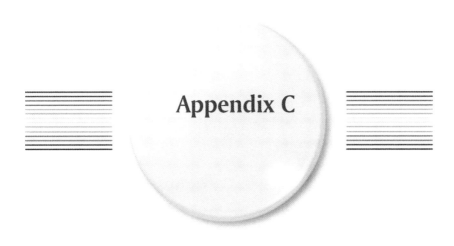

Appendix C

Selected Bibliography

The following books give a wonderful insight into the early years of spectroscopic development. These are available for download from http://www.archive.org/.

Browning, J. *How to work with the Spectroscope*. Privately published by Browning (1882).

Maunder, E. W., Sir *William Huggins and Spectroscopic Astronomy*. TC & EC Jack (1913).

Norman Lockyer, J. *The Spectroscope and Its Applications*. McMillan & Co (1873).

Proctor, R. A. *The Spectroscope and Its Work*. Society for promoting Christian knowledge (1888).

Olivier Thizy (Shelyak) has also prepared a "recommended" reading list:
http://www.shelyak.com/en/links.html

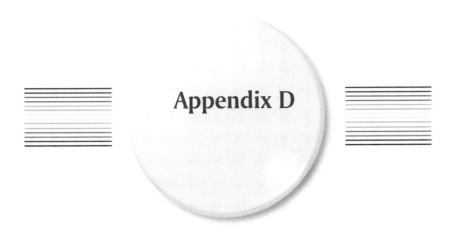

Appendix D

Extras

Springer has provided a web page to host a collection of additional material that compliments this book.

All the weblinks listed in the book are available as well as color illustrations of various spectra and photographs that are only printed in black and white in the book.

Excel spreadsheets are also available to assist in the design of spectroscopes:

- TransSpec.xls provides all the necessary calculations to evaluate a transmission grating in a converging beam.
- SimSpec.xls is a spreadsheet originally prepared by Christian Buil and translated into English (with annotated comments) to assist in the design of a transmission/reflection grating spectroscope.

These materials can be accessed through:

http://extras.springer.com/

Springer customers can search the extra materials by ISBN (please make sure you enter the full ISBN number, including hyphens).

A support group dedicated to assisting amateurs in astronomical spectroscopy has been set up by the author:

http://tech.groups.yahoo.com/group/astronomical_spectroscopy/?yguid=322612425

Membership of this group is open to any interested amateur. The group provides an ideal forum in which to raise questions on any and all aspects of spectroscopy.

There are many files available in the "knowledge base" covering spectroscope design and construction as well as the practical aspects of spectral imaging and subsequent analysis. Spectrum images are regularly uploaded to assist the novice, and there are many experienced amateurs who are only too willing to discuss any problem you may have as you move up the learning curve.

Any corrections and or amendments to this book will be published on the Astronomical Spectroscopy forum. See you there. . . . . .

Index